THE
WONDER
OF
MATHEMATICS

THE
WONDER
OF
MATHEMATICS

Suleiman Gierdien

Library of Congress Control Number:		2010906919
ISBN:	Hardcover	978-1-4500-9232-6
	Softcover	978-1-4500-9231-9
	Ebook	978-1-4500-9233-3

To order additional copies of this book, contact:
Xlibris Corporation
0-800-644-6988
www.xlibrispublishing.co.uk
orders@xlibrispublishing.co.uk
300227

CONTENTS

INTRODUCTION

All fundamental principles and concepts of mathematics can be adequately explained in *real life* through showing how real things work or by analyzing the underlying principles in real terms. In mathematics the least effective way of teaching concepts and ideas is to encourage memorization of rules and formulae like a parrot, to the extent that if we forget the rule or formula we are stuck. *Logical* reasoning and comprehension of underlying principles in formulae should be promoted as the key to solving any problem. To implement and facilitate this shift in the mentality of our learners and even our teachers, we need to embrace an old concept that got lost among the multitude of modern-day rules and regulations, which govern the way we perceive mathematics:

We should *see* mathematics, not merely *think* mathematics.

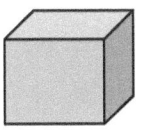

UNDERSTANDING BASIC ALGEBRA

Multiplication in the Real World

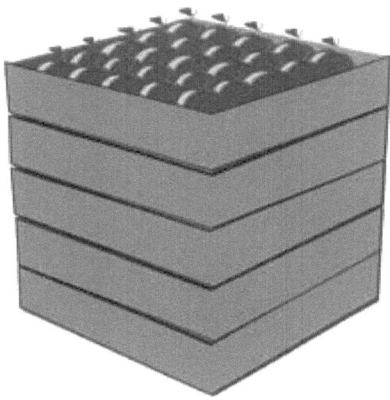

There are 5 rows of tomatoes both ways per box, and there are 5 boxes. How many tomatoes altogether?

$$5 \times 5 \times 5 = 125$$

Did you notice how we couldn't even see the tomatoes in the other 4 boxes and how we didn't need to count the tomatoes in the first box one by one?

That was because we were calculating in 3-D because in the real world *multiplication is fundamentally a 3-D operation.* That concept is the key to transforming our mind-set toward algebra as we know it.

What Is Algebra?

Algebra is a fundamental area of mathematics that entails the practice of substituting numbers by symbols, to show that certain rules apply to any real numbers and/or operations.

One of the greatest misconceptions about algebra is that it is only about numbers, symbols, and linear equations, instead of being *seen* as the gateway into the astounding *3-D world of mathematics*. Algebra shows patterns in real life, real life we can *see*, so why not algebra?

Understanding Linear, 2-D, and 3-D

Linear:

o A line is linear, i.e., has one *single dimension*, like measurements on a ruler. Writing in a book is linear, i.e., in a straight line. A linear equation is a straight-line equation.

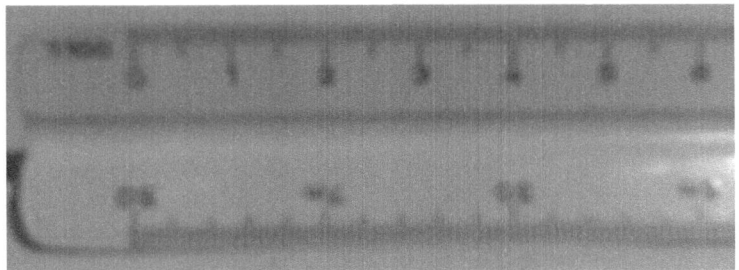

2-D (2-dimensional):

o This is a *planar* object, i.e., a flat plane with length and breadth, but no height or depth. In simple terms, two dimensions perpendicular (90°) to each other.

Breadth

Length

3-D (3-dimensional):

    o   This is a *box-shaped* object, i.e., has *length, breadth, and height,* simply three dimensions perpendicular to each other; this is how we perceive objects in the real world. In 3-D presentations, like on paper, we usually see two sides and either top or bottom of the object, which is meant to represent real-life 3-D objects where we can walk around the object and see underneath and on top.

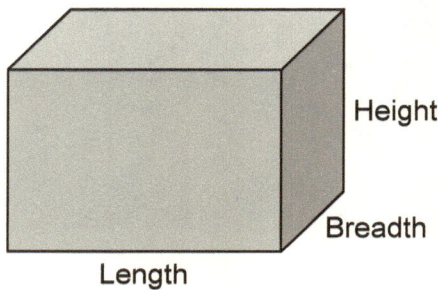

3-D block with hidden lines

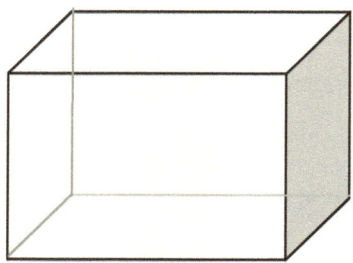

3-D wire-frame block

Basic Calculations – Multiplication

$$a \times a = aa = a^2$$

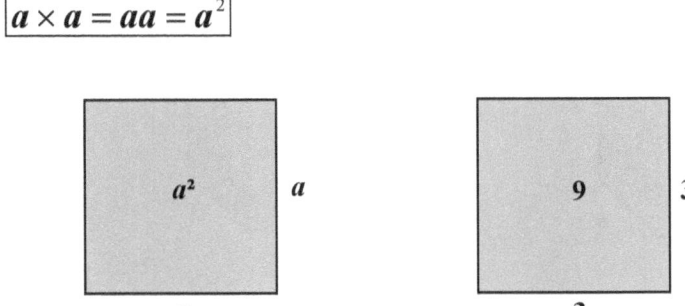

See how a x $a = aa = a^2$ (a squared), same as real numbers 3 x $3 = 3^2 = 9$. Do you see the logic, how a^2 is not just an a with a little ^{2} on the top (also called "to the power 2") but a square 2-D block with a value of a^2. If we wanted to clearly show that the real numbers represent a square block, we could have said the block 3^2 as well.

$$a \times b = ab$$

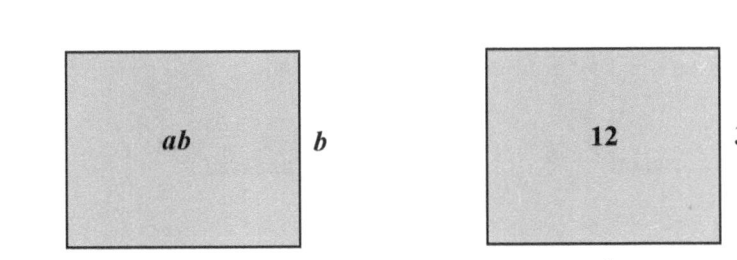

See how a x $b = ab$, like real numbers 4 x $3 = 12$. Here we just group the numbers in a cluster, ab representing the block a x b.

$$a \times a \times a = aaa = a^3$$

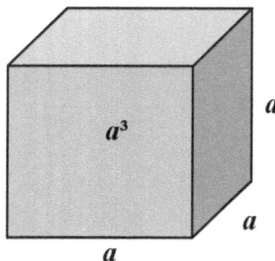 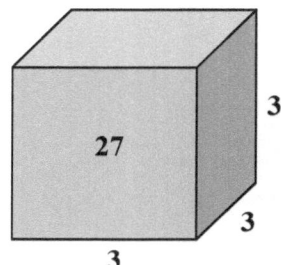

See how a x a x $a = a^3$ (a cubed), representing a cube (3-D shape with equal sides all round, which from now on we will simply call a *3-D block*). Remember, a^3 represents the whole block.

$$a \times b \times c = abc$$

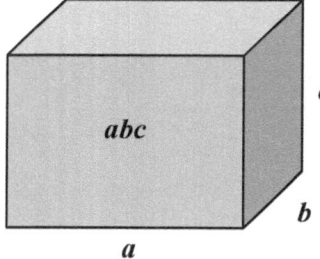 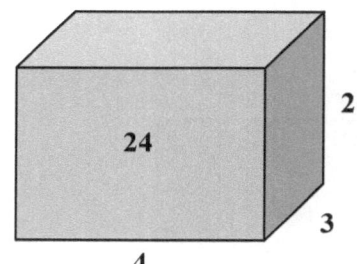

See how this rectangular 3-D block has different dimensions to each side.

Basic Calculations – Roots

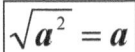

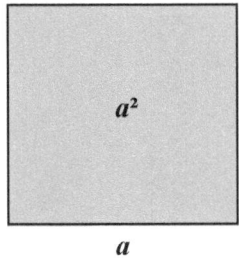

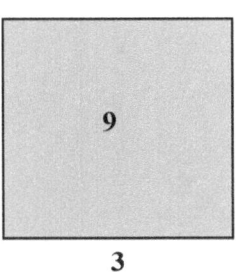

$\sqrt{}$ (square root) means which number multiplied by itself will give us the number in question. $\sqrt{a^2} = a$, because $a \times a = a^2$.

$$\sqrt[3]{a^3} = a$$

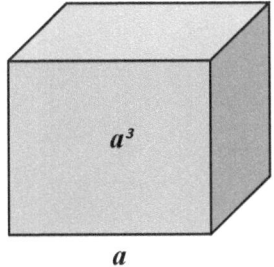

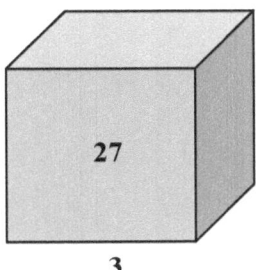

$\sqrt[3]{}$ (cube root) means which number multiplied by itself 3 times will give us the number in question. $\sqrt[3]{a^3} = a$, because $a \times a \times a = a^3$.

Understanding 4th Dimension

Let's use real numbers to better understand this concept:

$$2 \times 2 \times 2 \times 2 = 2^4 = 16$$

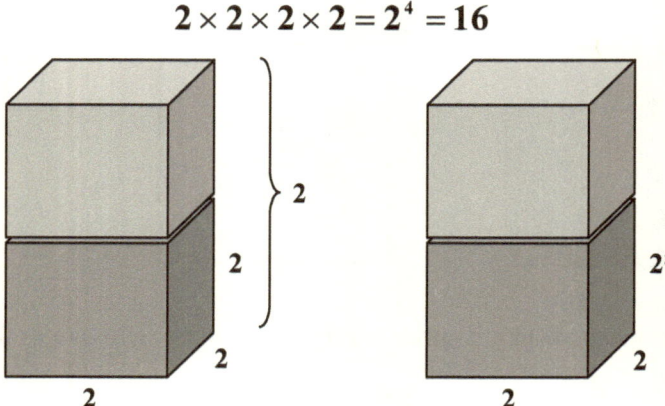

See how the first 3 dimensions create a 3-D block, then the 4th dimension doubles that block.

That was the number 2^4 ; let's now see how the number 3^4 behave:

$$3 \times 3 \times 3 \times 3 = 3^4 = 81$$

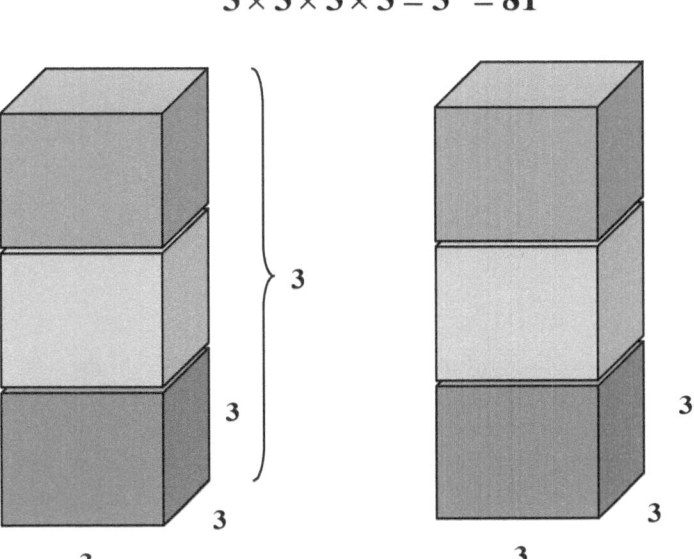

Aha! See how the first 3 dimensions create a 3-D block, then the 4th dimension triples that block. So what have we learned from this? That the 4th dimension is a *multiple* of the 3-dimensional block.

To experiment with other dimensions on your own, take $0.01 and then double it every day for 30 days. That will give a pretty good idea of how multiplication works in other dimensions, the last dimension doubling the previous one. That will be the same as $0.01 x 2 to the power 30 ($0.01 x 2^{30}).

$$a \times a \times a \times a = aaaa = a^4$$

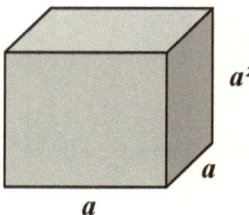

a^2

a

a

For practical reasons we will adopt the number 1 as the 4th dimension for graphical purposes since the 4th dimension is undefined; what that means is that we will draw basically the same size block to represent 3- and 4-dimensional blocks, the 4th dimension will then only be expressed as a multiple of the 3rd.

$$a \times b \times c \times d = abcd$$

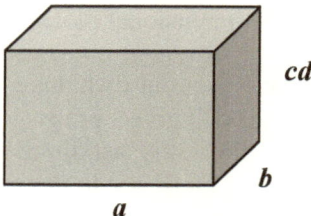

cd

b

a

Basic Calculations – Division

$$\boxed{\frac{a \times b \times c}{d} = \frac{abc}{d}}$$

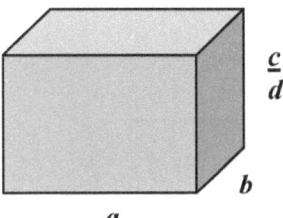

See how height is split into a fraction? If d was 2, the height would be half, or a third for 3, etc. Let's see what this would look like graphically:

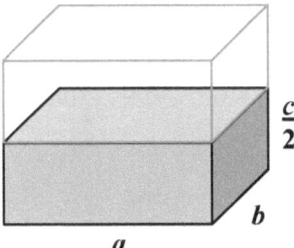

We could also have divided by any of the other sides that would have yielded the same result, but can you see what that would imply? That we would split the block through vertically instead of horizontally; it wouldn't matter though.

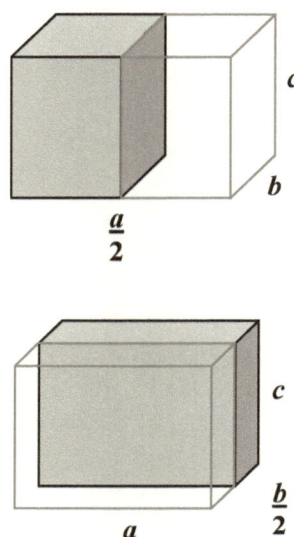

Do you see how division happens in any single dimension, even though we meant to divide the whole block? Whichever dimension we choose to divide in does not change the answer, but do you *see* the logic in the blocks?

The Underlying Principles

The most important concept we have learned thus far in our 3-D algebra is that *multiplication and division create and modify* blocks, each component being integral to the structure of the block. All components of blocks are created perpendicular (90°) to each other. The general direction of block creation is up. There are no limits to the amount of dimensions in one operation, thus creating virtual columns potentially going up to infinity. Multiplication and division are essentially *3-D operations*.

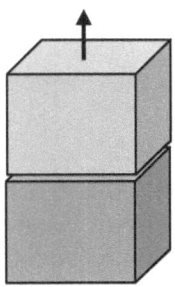

Addition and subtraction, on the other hand, do not create or modify blocks but merely *group or ungroup* them on the *2-D plane*, making addition and subtraction essentially a *2-D operation*. Addition and subtraction do not recognize the integral structural components of blocks as separate entities but see a block as a whole entity on its own, much the same as we would see an apple as an entity on its own, ignoring the internal core and other parts as separate entities.

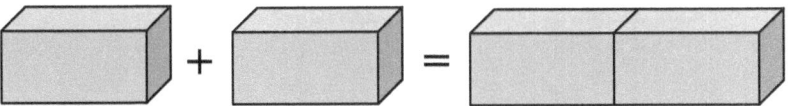

Basic Calculations – Addition and Subtraction

$$A + A + A + B = 3A + B$$

Addition and subtraction only see blocks as alike or not alike and group them accordingly, like apples will be grouped with apples and oranges with oranges.

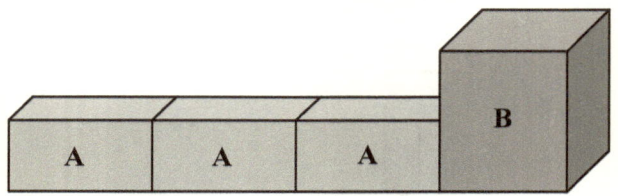

The illustration shows the basic arrangement of an addition operation. See how addition groups all the blocks together, but like and unlike *blocks* will *maintain their own identities* within the resultant answer.

$$3A - A + B = 2A + B$$

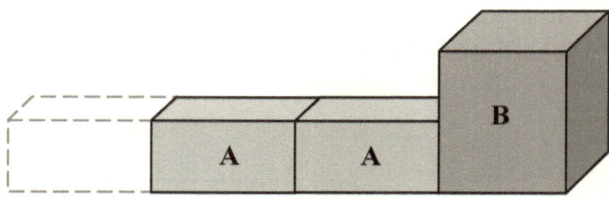

See how subtraction just eliminates entities from the selection, basically just ungrouping them.

Basic Calculations – Factorization

$$\boxed{abc + dbc = (a + d)bc}$$

Let's see how factorization works. As we can see, the two blocks have the same dimensions on two sides, so if we connect the blocks, we will have a perfect butt joint.

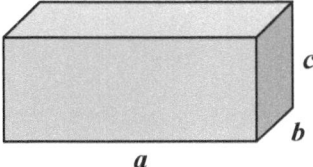 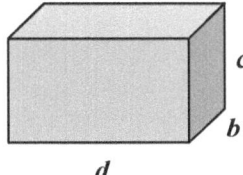

Resultant block:

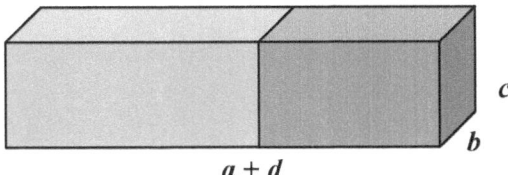

See how this $2bc$ $(a + d)$ would simply mean 2 of those combined blocks!

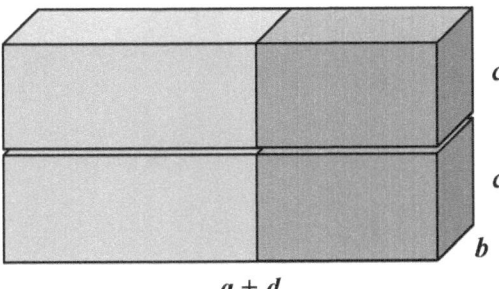

$$(a + b)(a + b) = a^2 + 2ab + b^2$$

This is one of those mysteries of mathematics that hardly anyone understands; we just know which numbers to multiply with which, but the logic? It's typical algebra, we say.

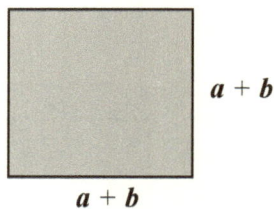

A 2-D block with multiple subcomponents for length and breadth, that's all it is.

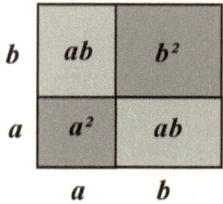

It's a simple multiplication exercise, creating blocks.
Shown below is the linear procedure for multiplication in this type of calculation:

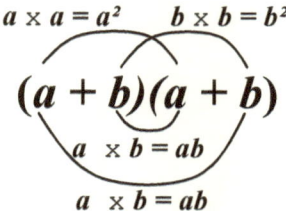

Let's add a few blocks so that we can see the pattern more clearly:

	a	a	b	b
b	ab	ab	b^2	b^2
b	ab	ab	b^2	b^2
a	a^2	a^2	ab	ab
a	a^2	a^2	ab	ab

See how each small block is created by multiplying the column value by the row value.

Row $\Longleftrightarrow$ a | ab |

b

$\Updownarrow$ **Column**

If we go back to our example and look at the values outside the main block, we should see instantly that the factors (sides of block) are:

$$(2a + 2b)(2a + 2b)$$

and the product (within the main block) is:

$$4a^2 + 8ab + 4b^2$$

Let's see what the same procedure would look like if we substitute real numbers.

3	6	6	9	9
3	6	6	9	9
2	4	4	6	6
2	4	4	6	6
	2	**2**	**3**	**3**

Thus:

(2x2+2x3)(2x2+2x3)

(4+6)(4+6)

= 16 + 48 + 36

= [4a^2 + 8ab + 4b^2]

And that's where we leave it in algebra till we substitute real numbers.

Fortunately, as soon as we work with real numbers, we go directly to the result:

(2x2+2x3)(2x2+2x3) = 10 x10 = 100

BODMAS

For want of a better way, this was how we were taught the *order of preference* in mathematical operations by BODMAS.

B   *Brackets 1st*

Over (takes preference over)

D
M   }   *Division and Multiplication 2nd*

A
S   }   *Addition and Subtraction Last*

What does this mean in real terms though, because BODMAS is not even a word. The logic is in the blocks:

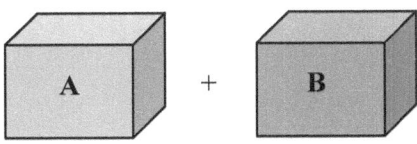

See how we are adding 2 blocks together? But what if we break these blocks into components?

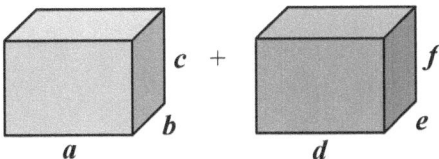

Let's use real numbers to explain this concept more efficiently:

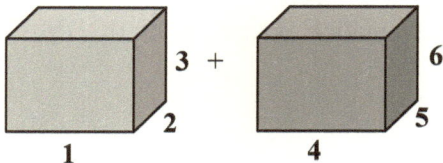

$$1 \times 2 \times 3 + 4 \times 5 \times 6$$

Can you see in this configuration how we can't do 3 + 4, or even 1 x 2 x 3 = 6, then 6 + 4 = 10, and then continue 10 x 5 x 6 = 300. That would go against the fundamental structure of the blocks, wouldn't it? So it makes perfect sense to do the multiplication 1st, then the addition, thus:

$$1 \times 2 \times 3 + 4 \times 5 \times 6 =$$
$$6 \quad + \quad 120 \quad = 126$$

Now where do brackets fit in?

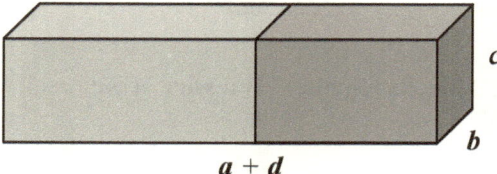

$a + d$

Remember our factorization exercise?
Let's substitute real numbers:

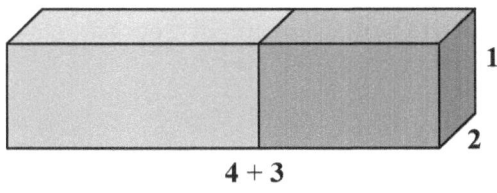

4 + 3

4 + 3 x 2 x 1

See how this is *not* a logical configuration? It means *4* as a separate entity +
the block *3 x 2 x 1*, doesn't it?

(4 + 3) x 2 x 1

This is where brackets come in to group integral parts of a component
together so that it may be given preferential treatment, which is the logical
configuration. Do you see how 4 + 3 needs to be solved first before the
multiplication can be tackled?

Understanding Multiplication with Negative Numbers

This is usually an area of mathematics where logic takes a leave of absence and we resort to the holy book of rules. In great detail we explain about the Cartesian coordinate system and how if we multiply, our answer will fall into one of the quadrants, either negative or positive.

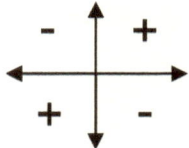

Then there are usually the summary of all this information:

$$(+) \times (+) = (+)$$
$$(-) \times (-) = (+)$$
$$(-) \times (+) = (-)$$
$$(+) \times (-) = (-)$$

And the conclusion:

When the signs are the same, answer will be positive; if not, answer will be negative.

But the why stays unanswered.

Let's shed some light on this by simple logic.

By Example of Deficit:

What does a negative number mean in real life? The simplest explanation would be that it is a debt, or units owed, or a *deficit*.

Example:

- We have no money in the bank, but on the last day of the month they deduct service charges from our account, $50. Leaving a balance of -50. See how we now have less than nothing, i.e., we owe. Next month if we don't put money into the account, that amount owing will *double*.

$$-50 \times 2 = -100$$

- -50 x -2 will be the opposite of that, i.e., a reversal of the debt to a gain.

$$-50 \times -2 = 100$$

Using the Logic of the Number 1:

If we multiply any number by the number 1, the answer will stay the same as the original number, right?
So it will be obvious that

$$-50 \times 1 = -50$$

So in turn, the opposite,

$$-50 \times -1 = 50$$

That's all; there are really just 2 cases, which logic will apply even if we turn the numbers or the signs around.

UNDERSTANDING EQUATIONS

What is an equation?

The fundamental structure of equations is that one side is always equal to the other, one side usually showing the structural components or elements, the other side showing the result, like in the following:

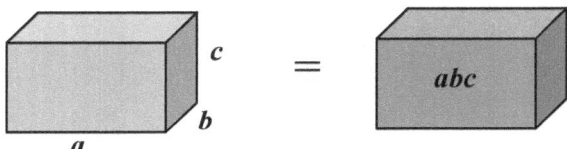

Do you see how the two sides of this equation are exactly the same?

Equations do not necessarily have to be in the above order, a x b x c = abc, but may be equated to something that does not reflect the structural components at all. Example:

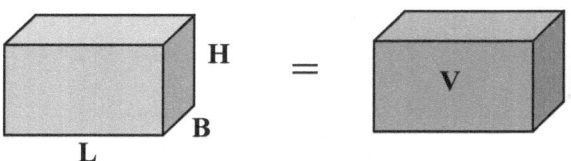

See how $L \times B \times H = V$

Manipulating Equations

Example 1: Multiplication and Division

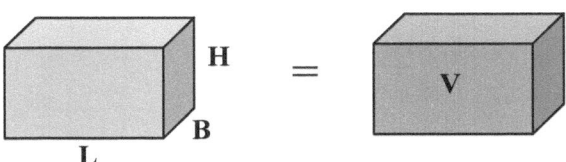

In the above equation we can clearly see that if we multiply L x B x H we will get the result V, but what if we have a predetermined value for V and one of the structural components is unknown? Let's see what it would look like with real values:

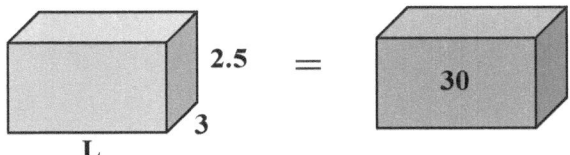

What would be the logical procedure to find L? If we could manipulate the equation so that L = ___, that would be the logical solution, wouldn't it?

Let's go back to the original equation now so that we can see exactly what it is we need to do:

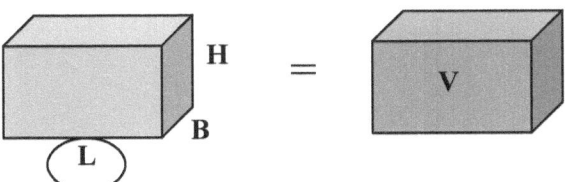

We will remember that multiplication is the operation that creates structures by grouping components into blocks; now we will need the inverse of that operation to dismantle the block, to get to component L. Division is the inverse of multiplication, thus:

$$\frac{L \times B \times H}{B \times H} = L$$

Now remember, what we do to one side we have to do to the other, relevant to what is there already, so:

$$\frac{L \times B \times H}{B \times H} = \frac{V}{B \times H}$$

See how we divided both sides by BH. Now see how $B \times H$ cancels each other out on the left side.

$$L = \frac{V}{BH}$$

$$L \qquad = \qquad \frac{V}{BH}$$

Can you see how both sides of the equation now refer to the linear component L?

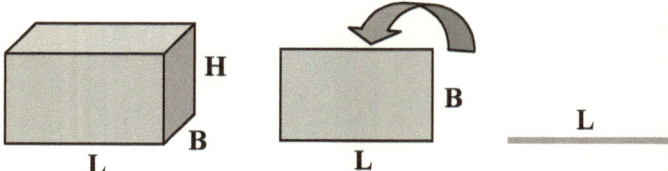

The illustration shows the systematic dismantling of components:

$$\left(\frac{L \times B \times H}{H} = L \times B \right) \quad \text{and} \quad \left(\frac{L \times B}{B} = L \right)$$

Okay, we've seen the logic. Now let's see what this equation will look like with the numbers

$$L \times 3 \times 2.5 = 30$$

$$L = \frac{30}{3 \times 2.5}$$

$$L = 4$$

We could've added the extra step:

$$\frac{L \times 3 \times 2.5}{3 \times 2.5} = \frac{30}{3 \times 2.5}$$

But can you see how they would just cancel each other out?

So basically the derived logic here will be that if we want to eliminate a number from above the line on one side, we just move it to the other side below the line. Can you see that?

$$\frac{L \times B \times H}{1} = \frac{V}{1}$$

$$L = \frac{V}{B \times H}$$

Example 2: Multiplication and Division

In this equation, h is the unknown:

$$\frac{h \times \pi}{3} = V$$

$$\frac{\pi \times r^2 \times \boxed{h}}{3} = V$$

We need to eliminate 3, so both sides x3:

$$h \times \pi = 3V$$

$$\pi \times r^2 \times h = 3V$$

Now we divide by πr^2.

$$h = \frac{3V}{\pi r^2}$$

$$h = \frac{3V}{\pi r^2}$$

Can you see how both sides of the equation now refer to linear component h? In this example we have seen how x is inverse to ÷ and vice versa.

Example 3: Square and Square Root

In this equation, C is the unknown:

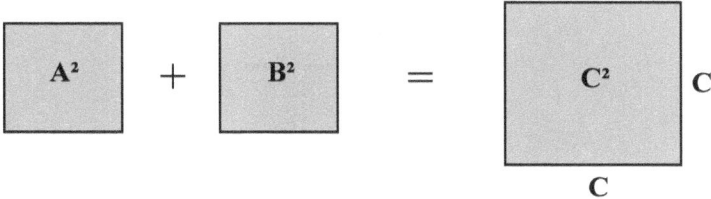

$$A^2 \quad + \quad B^2 \quad = \quad C^2 \quad C$$

$$C$$

Thus, the following conclusion can be deduced from that:

$$A^2 + B^2 \quad = \quad C^2$$

$$C \qquad\qquad\quad C$$

Now we want C, not C^2. Remember how square root means, which number multiplied by itself will give the number in question?

$$\underline{C} \qquad = \qquad \underline{\sqrt{A^2 + B^2}}$$

In this example we have seen how $\sqrt{}$ is inverse to ^{2}, hence also vice versa.

Example 4: Addition and Subtraction

Addition and subtraction are by far the easiest to manipulate, the one being the inverse of the other.

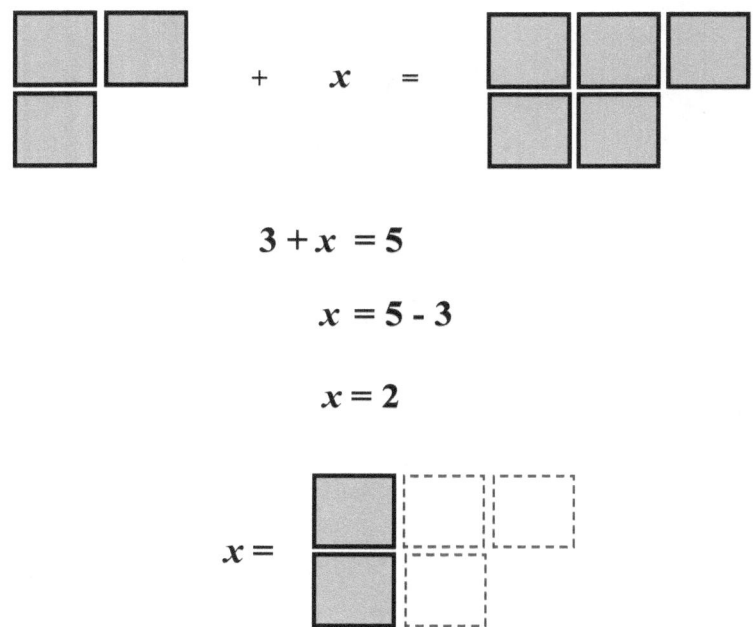

$$3 + x = 5$$

$$x = 5 - 3$$

$$x = 2$$

See how 3 becomes -3 on the other side of this equation, meaning this:

$$3 - 3 + x = 5 - 3$$

Do you see how $3 - 3 = 0$, i.e., cancels each other out?

Example 4: Multiple Blocks

$$V = A \times B \times \textcircled{C} + D \times E \times F$$

C is the unknown in this equation.

See how we need to eliminate the other block first before we can dismantle block ABC.

$$V - DEF = ABC$$

Now we can dismantle the block:

$$\frac{V - DEF}{AB} = C$$

$$\left| \quad \frac{V - DEF}{AB} \quad = \quad C \quad \right|$$

Example 5: Factors

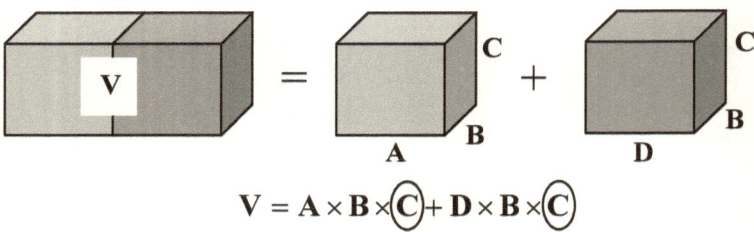

$$V = A \times B \times \textcircled{C} + D \times B \times \textcircled{C}$$

C is the unknown in this equation! Now how do we handle that?

Logic, guys, we need to factorize:

$$V = BC\,(A + D)$$

$$\frac{V}{B\,(A + D)} = C$$

$$\left| \frac{V}{B\,(A + D)} = C \right|$$

Terms, the Linear Name for Blocks

Some of us may ask the question, why were we never explained the logic of the blocks of algebra?

Surprisingly enough, we were, but the name was just unrecognizable and incompatible with 3-D, because we learned the linear name: terms.

Terms are separated by + and - . See how we are in fact referring to the blocks of algebra as a series of numbers.

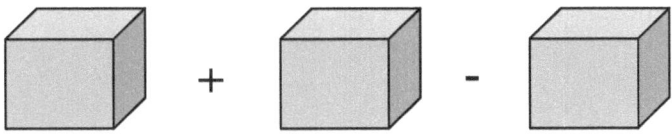

That is the reason why the vast majority of learners find it extremely difficult to learn mathematics, because in our efforts to simplify mathematics, we haven't taken into consideration that to fully comprehend *linearization* in mathematics we have to understand the *3-dimensionality* first. That is the gift all of us were born with: *to see things and instantly recall the pictures in our minds.*

It all starts with the basics, the *blocks* of algebra, which hold the *key* to comprehension of all the other fields within mathematics. Throughout this book we will see how understanding the blocks in algebra does contribute to in-depth comprehension of formulae that we have taken for granted for so long without having a clue as to the simple fundamentals behind them.

CALCULATIONS FOR BOX SHAPES

Perimeter for Box Shapes

Calculating perimeters (outer edges) on straight-line shapes is probably the easiest calculations we will ever do because it's basically just *adding* the sides of a shape together and getting a result.

Square (All Sides Equal)

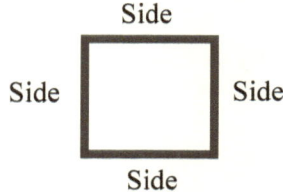

Perimeter = Side x 4

Rectangle (2 Opposite Sides Equal)

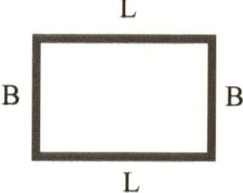

Perimeter = 2L + 2B
(Twice Length + Twice Breadth)

Triangle (3 Sides, Not Necessarily Equal)

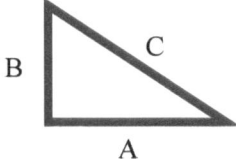

Perimeter = A + B + C
(Add 3 sides together)

Hexagon (6 Sides Equal)

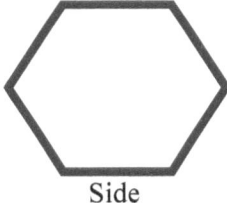

Side

Perimeter = Side x 6

Perimeter calculations are necessary for many applications, e.g., what length of timber molding will go around a picture frame, or how many meters of rope will go around a cricket field, or how many meters of gutters will go around a house. If we look around us, we will see countless examples of perimeters.

Area for Box Shapes

Surface area deals with flat surfaces (2-D), for instance, how many square meters of tiles or carpets would be needed for a room, etc.

3 Golden Rules:

1. Length times breadth (L x B) is always the basis for *any* area calculation.

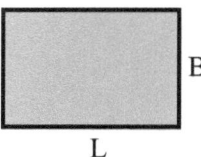

❖ Note: Area is predominantly measured in m² (square meters); that is already a clue as to the calculation
   m x m = m².

2. Length and breadth dimensions are always measured *perpendicular* (90°) to each other.

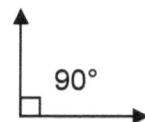

3. Any shape that is not a square/rectangle can be equated to a fraction / multiples of a square / rectangle by using a *ratio*.

Right-angled Triangle = Fraction of Rectangle

Nonright-angled triangle = Fraction of Rectangle

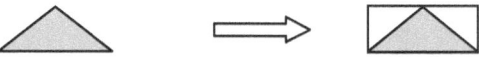

Trapezoid = Average Rectangle

Circle = Multiple Squares

(For the sceptics: Yes! These rules apply for circular shapes as well; they will be discussed in detail in next chapter "Calculations with Circular Shapes.")

Square

$$\text{Area} = \text{side}^2 \text{ (side squared or side x side,}$$
could well be $L \times B$, same as rectangle)

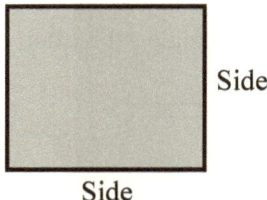

Side

Side

Example: If square is $2m \times 2m$,
$$\text{area} = 2m \times 2m = 4m^2$$

Rectangle

$$\text{Area} = L \times B \text{ (Length x Breadth)}$$

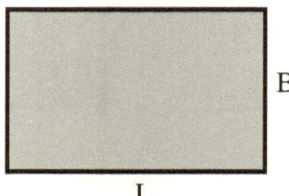

B

L

Example: If rectangle is $4m \times 3m$,
$$\text{area} = 4m \times 3m = 12m^2$$

Doesn't this look exactly like the blocks in algebra?
Yes, it does because it is exactly the same concept; this is how the multiplication pattern works, just here we calculate real planes or real blocks, while in algebra we create virtual planes or blocks (according to the multiplication pattern).

Right-Angled Triangle

$$\text{Area} = \frac{L \times B}{2} \quad \text{(Length x Breadth divided by 2)}$$

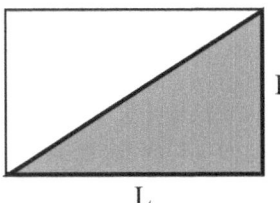

B

L

See how 2 identical triangles fit into rectangle.
(Meaning one triangle is half of rectangle)

Example: If sides of triangle are 4m x 3m,
$$\text{area} = \frac{4m \times 3m}{2} = 6m^2$$

Nonright-Angled Triangle

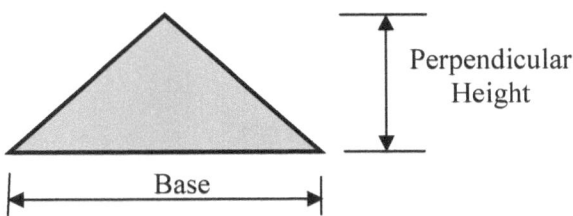

Perpendicular
Height

Base

½ Base x Perpendicular Height

This is the conventional way we learned this formula, but if we look at the golden rules this is implied for any shape anyway, and besides, it would defy logic if we suddenly decided to use the diagonal dimension, wouldn't it?

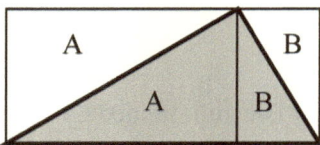

Can you see the logic? The A triangles are identical and so are the B triangles. We have an A and B triangle inside the original triangle, and the remainder of rectangle has an A and B. See, a nonright-angled triangle is *half of rectangle*, same as right-angled triangle, so why complicate matters, same logic, same formula.

$$\frac{L \times B}{2}$$

Trapezoid (Shape with 2 Parallel and 2 Nonparallel Sides)

By just looking at this shape, it's obvious that we could break it up into 2 triangles and a rectangle, right?

Thus:

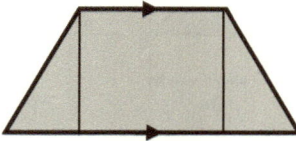

Now, we just calculate area of rectangle + area of the 2 triangles (or triangles individually if they are not identical).

But there is an easier way: if we measure on the center line as shown below we could *convert this shape to a rectangle* in an instant.

See how the A triangles are identical and so are the B triangles. That means what we cut away from the trapezoid we just add to the rectangle.

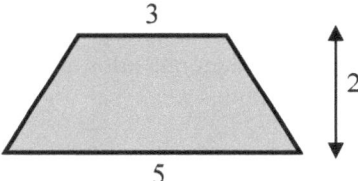

Now, how will we configure the length of the rectangle?
If we add 3 and 5, then divide the sum by 2, we will have the *average*:

$$\frac{3+5}{2} = \frac{8}{2} = 4$$

There we have it! Simple, isn't it? It we use elementary common sense, the formula will be:

$$\text{Area} = \frac{\text{Sum of 2 parallel sides}}{2} \times \text{Height}$$

Meaning:

Area = Average Length x Breadth

Parallelogram (Opposite Sides Parallel But Not Perpendicular)

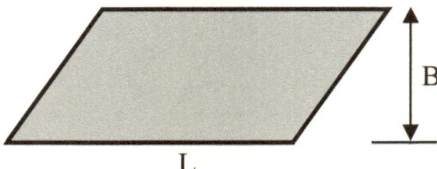

This should be pretty obvious after the trapezoid exercise, shouldn't it? Surprise! It's easier; because opposite sides are the same, no average required, unless, of course, if the diagonal sides are not parallel, then we would need to configure the average.

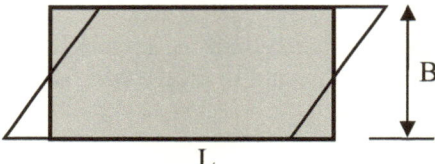

See how the pieces of triangles will fill up the rectangle. So simply:

$$\text{Area} = L \times B$$

By now we should have a fairly good understanding of how the golden rules work for areas with box shapes. Let's move on now to volumes for box shapes.

Volumes for Box Shapes

Rectangular Boxes

Volume is the capacity or space inside an object, measured in cubic meters (m^3), like quantity of concrete required for a foundation or the unit of quantities loaded by tipper trucks.

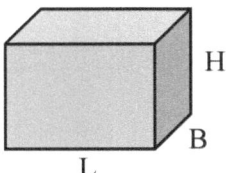

Volume calculations are merely an extrusion (stretching out into 3-D space) of surface areas, areas being 2-dimensional (flat) objects.
If we would start from scratch, the formula for volume would be:

$$V = L \times B \times H$$
(Volume = Length x Breadth x Height)

Or if area had already been configured:

$$\text{Volume} = \text{Area} \times \text{Height}$$

❖ Note: The unit for volume, m^3 (cubic meters), is already a clue as to the calculation:

$$m \times m \times m = m^3$$
$$m^2 \times m = m^3$$

Pyramids

Sometimes 3-D objects can be complex compositions of objects modified or even tapered in different planes.

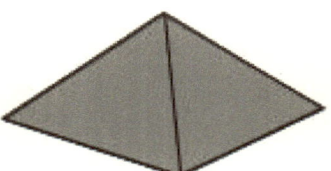 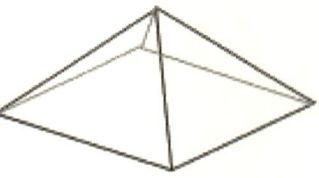

A pyramid, for example, is a shape with a square/rectangular *base that converges to a point* above. So how will we calculate volume for such a complex object? It is quite simple really; we need to know what the *ratio* is between a pyramid and rectangular block of the same size, like how we configure area of triangle to rectangle.

Let's see if we can configure the ratio ourselves. We just need a couple of identical pyramids that all slope 45° for this exercise.

Let's take one pyramid and place it with base flat on the ground, then turn 4 pyramids on their sides as shown.

Right, now we just have to fill the gap with the last pyramid upside down.

Okay, now let's count how many pyramids closed this block. Six! But let's analyze before we get excited.

What we want to know is, *what is the ratio between a pyramid and a rectangular block of the same base and height?* Now, remember how we placed one pyramid upside down on top of the 1st pyramid? Can you see how we created a block of the same base but twice the height? That means we have 6 pyramids in 2 blocks, doesn't it? This in turn means 1 block should have 3 pyramids. Let's see, if we cut these positioned pyramids as shown:

The result would be:

4 half pyramids that equal 2 full ones, plus the one right in the center, which is currently hidden from view.

Thus pyramid equals ⅓ of rectangular block, hence the formula for volume of pyramid:

$$\text{Volume} = \frac{L \times B \times H}{3}$$

CALCULATIONS FOR CIRCULAR SHAPES

Circumference (Circle Perimeter)

What Is π (Pi)? (The Circumference Definition)

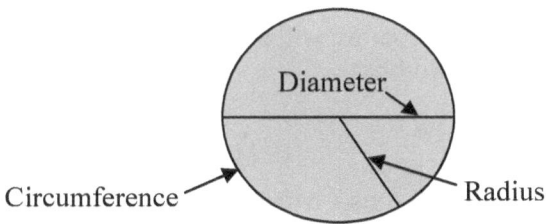

π is the ratio between the diameter and circumference of a circle and represents a number, approximated to 3.142 or $\frac{22}{7}$; that is the amount of times the diameter will go into the circumference of a circle.

If we straighten out the outer edge (circumference) of a circle, it would look like this compared to diameter:

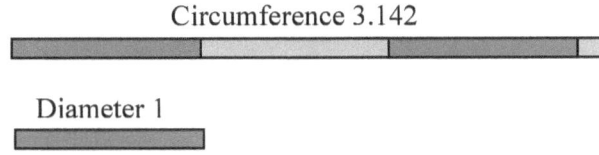

Hence the ratio 3.142 to 1.
This gives us the formula for circumference of circle:

Circumference = π x Diameter   [**C** = π**d**]

Let's see how that formula can be used to configure perimeter of an elliptical (oval) shape:

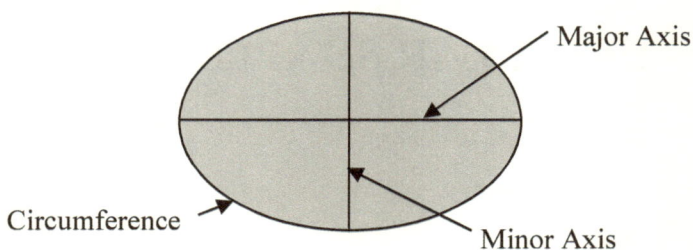

See how we now have two diameter lines of different lengths, but if we remember our trapezoid configuration, we will see in an instant that we need an average between these two lines.

Hence:

$$\text{Circumference} = \frac{\text{Major Axis} + \text{Minor Axis}}{2} \times \pi$$

Or: Circumference = π x Average Diameter

See, same basis as for circle: πd.

Area for 2-D Circular Shapes

Most of the time, when you ask someone who has left school as recently as two years ago how do you calculate area of a circle, they reply, "Pi something."

Why is that? Probably because they have memorized the formula like the vast majority of us, without comprehending the logic behind it; that is the trouble with memory: we tend to forget *abstract philosophy* very quickly.

Just to emphasize the pattern of all area calculations, let's quickly do a recap of the golden rules for area calculation:

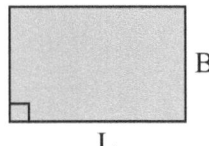

1. Length times breadth (L x B) is always the basis for any area calculation.
2. Length and breadth dimensions are always measured perpendicular (90°) to each other.
3. Any shape that is not a square/rectangle can be equated to a fraction / multiples of a square / rectangle by using a *ratio*.

Right, now let's see exactly how these rules apply to circles.

What Is π (Pi)? (The Area Definition)

Funny how we are asking the same question again, right? But there really are two definitions, one for *circumference* (the conventional definition) and one for *area* (the "hidden" definition).

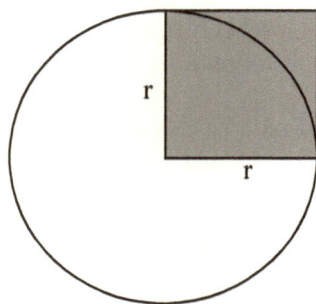

In the illustration, π is the number of times the surface area of the square will go into the circle. The squares below illustrate this ratio graphically:

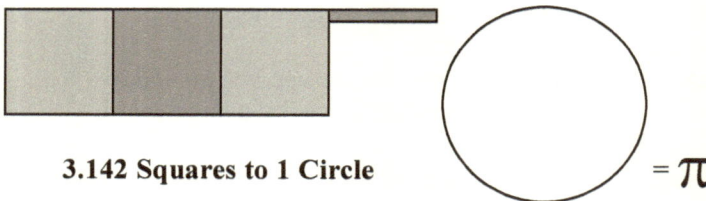

3.142 Squares to 1 Circle = π

We will learn more about π in the section "Regular Polygons and Circles." Now, let's analyze the formula for area of circle.

Area of Circle

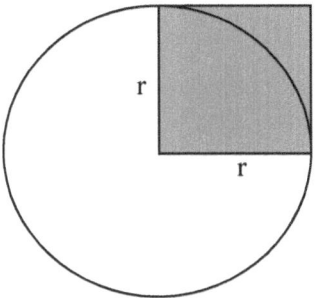

$$\text{Area of Circle} = \pi \, r^2$$

Same as:

$$\text{Area of Circle} = \pi \times r \times r$$

See how (r x r) is actually (L x B) of the square, and *π represents the ratio* of how many times the square goes into the circle.

Let's compare this to area of a triangle:

$$\text{Area of triangle} = \frac{L \times B}{2}$$

Same as: $\qquad \text{Area of triangle} = 0.5 \times (L \times B)$

$$\text{Area of Circle} = 3.142 \times r \times r$$

Implied:

$$\text{Area of Circle} = 3.142 \times (L \times B)$$

Isn't that exactly the same principle? That is what makes mathematics amazing, finding the similarities in things that are apparently not alike.

Area of Ellipse

Let's see how an ellipse confirms our L x B concept even better:

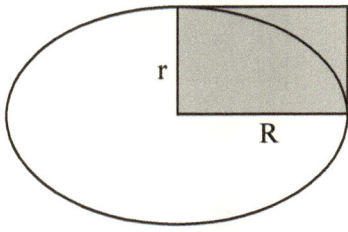

Area of Ellipse = π R r

Area of Ellipse = π x R x r

Area of Ellipse = 3.142 x L x B

Can you *see* how a circle and an ellipse have exactly the same logic? That is the how the patterns work in mathematics; if we know to which group or pattern an object or operation belongs, we will see the logic in the wink of an eye.

Working with Sectors

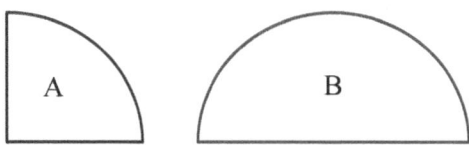

A is clearly a quarter circle, and B is a half circle. To configure area of these two should be quite easy, right? Area of quarter circle will be area of full circle divided by 4 or times 0.25, and half circle will be divided by 2 or times 0.5.

Now, see this funny little *sector*? How will we calculate area of that?

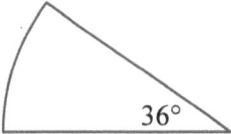

Okay, let's look at the logic. What is a quarter circle? Full circle being 360° and quarter circle 90°:

$$\frac{90}{360} = \frac{1}{4} = 0.25$$

Likewise for the little sector:

$$\frac{36}{360} = \frac{1}{10} = 0.1$$

Working with Segments

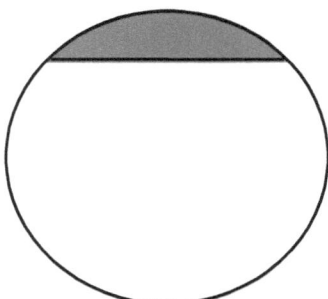

The shaded portion in the illustration represents a segment of the circle. So how do we calculate area for that?

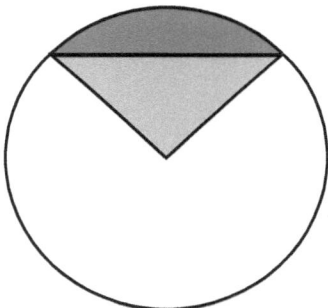

We know how sectors work, right? So we simply configure *area of the sector*, and then *deduct the triangle*. That will give us area of segment. We will see how this works later in the book.

That concludes our lesson on area for 2-D circular shapes for now.

Volume for Circular Shapes

Cylinder

As we have said before, 3-D objects are basically 2-D objects *extruded* into 3-D space, like a cylinder is a circle given *height/depth*.

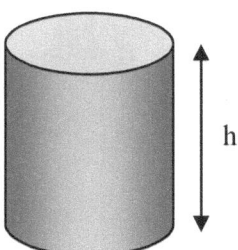

Formula:

$$\text{Volume of Cylinder} = \pi\, r^2\, h$$

Or:

$$\text{Volume of Cylinder} = \text{Area of Circle} \times \text{height}$$

That should be self-explanatory, so let's move on now to a more complex object.

Cone

We will remember that all box-shaped objects as discussed previously was either a square/rectangular block, or equated to a portion / multiple of a square / rectangular block.

Now, a cone should obviously be *equated to a portion of a cylinder* as that is the baseline template. That makes sense, doesn't it?

Now, if it was up to us to find this *ratio* between a cone and a cylinder, how would we go about it?

Just remember, we cannot just take the *average* radius from top to bottom of the cone, because that would just give us a cylinder with half the radius of the original cylinder, and if we inspect this shape closely, that *wouldn't do justice* to the cone, bottom having much more substance than the top.

Okay, let's take a look at what we know:

Volume of Cylinder = π r² h

Basically, that is all we have to work with, so what if we stack a series of flat cylinders onto each other with *reducing radii* on each? That would give us a rough idea of the ratio, wouldn't it?

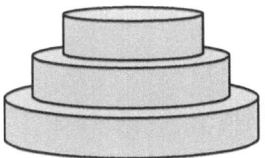

Basically, what we want to achieve is this:

But because we have no formula for tapering cylinders yet, we will assume the *average*, so that this,

will be *converted* to this, when measuring on the dotted center line.

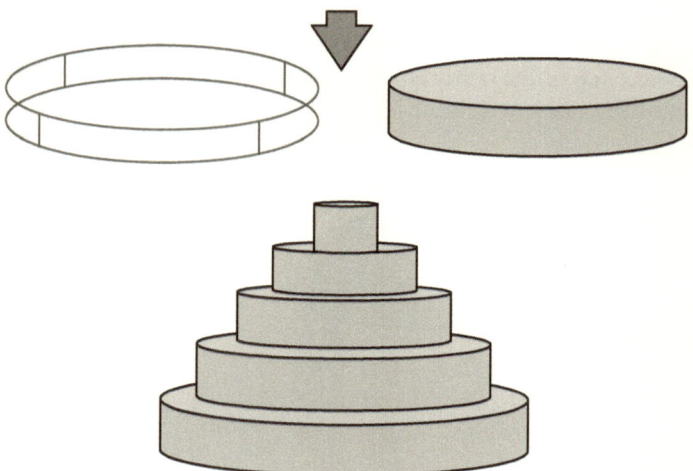

Obviously we cannot just use random cylinders; we have to use some logic here, starting with the *lines* of the cone and cylinder in *2-dimensional* side view.

Radius

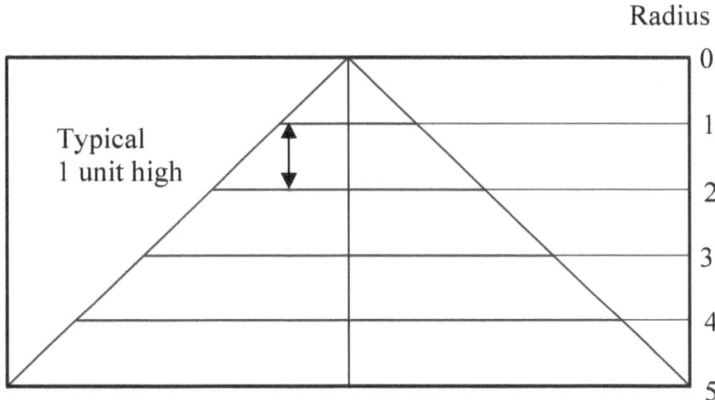

Then superimposing cylinders on them,

Average Radius

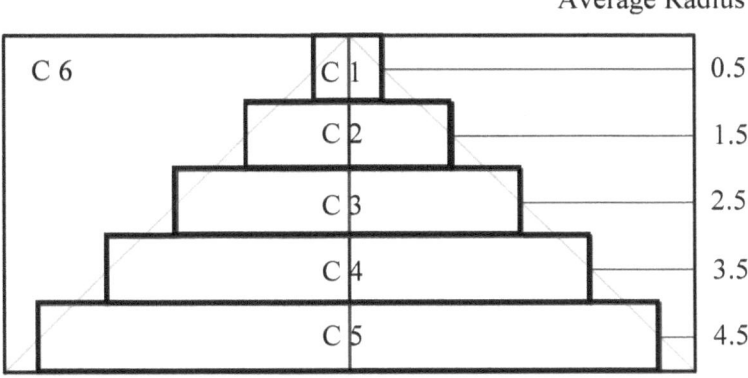

Now we calculate each segment of cone separately and add them together. That sum will be divided by the big cylinder.

$$\text{Ratio} = \frac{C1 + C2 + C3 + C4 + C5}{C6}$$

That looks like a lot of calculation, doesn't it? But not really, let's see how we can simplify this procedure by looking at one of the cylinders:

$$C1 = \pi \times r^2 \times h$$
$$C1 = \pi \times 0.5^2 \times 1$$

See how h = 1 won't change the result of this equation, so let's leave height out of it, thus:

$$C1 = \pi \times 0.5^2$$

$$\text{Ratio} = \frac{\pi \times 0.5^2 + \pi \times 1.5^2 + \pi \times 2.5^2 + \pi \times 3.5^2 + \pi \times 4.5^2}{\pi \times 5^2 \times 5}$$

Now see how π is a common factor in all the cylinder segments, so let's factorize:

$$\text{Ratio} = \frac{\pi (0.5^2 + 1.5^2 + 2.5^2 + 3.5^2 + 4.5^2)}{\pi \times 5^2 \times 5}$$

See how simple this calculation is now, one user-friendly simultaneous equation, instant result.

Result = 0.33 instead of 0.333333333. Great stuff! See how close we came to the accurate answer, with only 5 segments to the cone.

 Volume of Cone $= \dfrac{\pi r^2 h}{3}$

Clarifying Factorization

For those who may have lifted an eyebrow at the factorization of the cylinders involving π in our derived formula:

$$\text{Ratio} = \frac{\pi \times 0.5^2 + \pi \times 1.5^2 + \pi \times 2.5^2 + \pi \times 3.5^2 + \pi \times 4.5^2}{\pi \times 5^2 \times 5}$$

$$\text{Ratio} = \frac{\pi\,(0.5^2 + 1.5^2 + 2.5^2 + 3.5^2 + 4.5^2)}{\pi \times 5^2 \times 5}$$

Remember how area of a circle is calculated by a square as shown below?

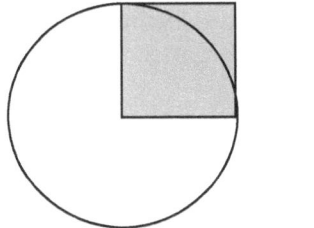 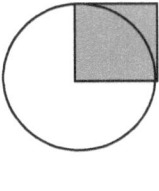

In our formula we simply added all those squares from the different circles together effectively *creating a big square equal to all the separate little squares together*. That result was then multiplied by π, to give us a *big circle equal to all the smaller ones*, which we would've created. In mathematics nothing just happens because of rules and regulations; everything has a logical explanation.

Can you *see* the 3-D algebra in this?

Sphere (Ball Shape)

This is the formula for calculating volume of a sphere:

$$V = \tfrac{4}{3}\,\pi\,r^3$$

The likelihood of anyone ever remembering such an obscure formula is remote to say the least, unless we can relate it to something concrete, which will hopefully stay with us long after we have forgotten the formula.

Let's start by breaking up this formula to see if we can find some logic hidden within.

$$\tfrac{4}{3} \times \boxed{\pi \times r \times r} \times r$$

Aha! We've found area of a circle.

$$\tfrac{4}{3} \times \boxed{\pi\,r^2 \times r}$$

Now, let's see, the only other formula that resembles this configuration is a cylinder (and what we learned in algebra is that the next dimension after planar is up into 3-D space):

$$\pi\,r^2 h$$

So, $h = r$, meaning, a cylinder where height equals radius!

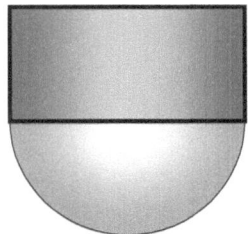

See what we have so far (side view), a cylinder half the height of the sphere. Let's continue:

$$\left(\tfrac{4}{3}\right)\pi\, r^3$$

Let's divide the cylinder into 3 parts, ⅓ each.

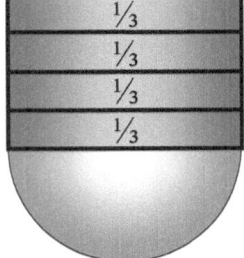

Plus another $\frac{1}{3}$, gives us $\frac{4}{3}$.

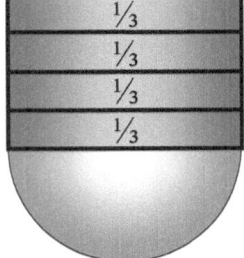

That's it guys, done it. We made sense out of the formula.

A sphere is calculated by equating it to a cylinder half the height of sphere, then increasing the height of cylinder by ⅓.

Isn't it funny how ⅓ keeps coming up when we deal with derivations from circles?

Cone = ⅓ of cylinder

Sphere = Half Cylinder + $\frac{1}{3}$ of cylinder

We could also have said a sphere is equal to a cylinder with a height $\frac{2}{3}$ the diameter of sphere. See if you can figure it out.

$$V = \tfrac{4}{3} \pi\, r^3 = \tfrac{2}{3} \pi\, r^2\, d$$

See how we combined the logic of algebra with circle geometry?

Okay, it seems we are on a roll here, so let's just look at one more formula before we leave this section on circular volumes.

Frustum of Cone (Cut-off Cone)

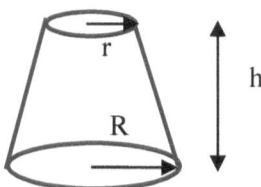

This is by far the most mysterious formula we will ever come across, not meaning the logic but the logic behind the logic.

This is the formula:

$$V = \frac{\pi h}{3}(R^2 + Rr + r^2)$$

Can you see the 3 entities in the formula?

$$V = \frac{\pi R^2 h}{3} + \frac{\pi R r h}{3} + \frac{\pi r^2 h}{3}$$

Volume = Big Cone + Elliptical Cone + Small Cone

Graphically it would look like this from above:

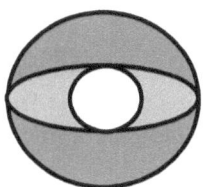

Meaning this, each shape a cone within the frustum.

Then we simply *add these shapes* together, and mysteriously we end up with the volume of the frustum.

Every concept in mathematics can be proven, so let's see. We need to construct a cylinder, then a cone within; then we cut the cone halfway, which will give us half the radius on top of frustum.

Then we calculate the frustum by the 3 entities as our analysis showed, where after we calculate the top part of our initial cone; the sum of that cone plus the frustum should be ⅓ of the cylinder.

Surprise! That's your homework. Take any height and any radius, but just remember, for this exercise the radius on top of frustum needs to be half of cylinder radius and frustum height half of cylinder. Congratulations to all of you for *see*ing this!

Just before we conclude this lesson on circular volumes let's just quickly look at another revelation arising out of the above.

Did you notice how a cone and a pyramid is both ⅓ of the entities in their respective groups, i.e., a cylinder and a box?

- So a frustum of pyramid should have exactly the same calculation as a frustum of cone; the logical pattern dictates that.

Volume of frustum of cone:

$$V = \frac{\pi h}{3}(R^2 + Rr + r^2)$$

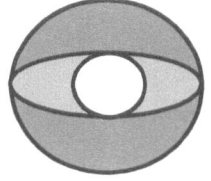 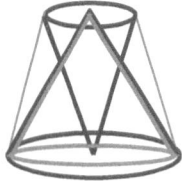

Volume of frustum of pyramid:

$$V = \frac{h}{3}(S^2 + Ss + s^2)$$

(Where S^2 is the base and s^2 is the top of frustum)

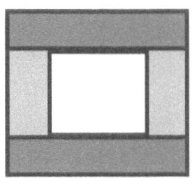 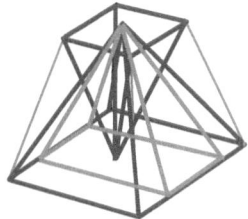

Example: Reservoir

Usually located on hilltops or the highest parts around towns or townships are reservoirs wherefrom water reaches our taps. The following illustration shows a typical reservoir. Dimensions are shown in 2-D section through. Let's see if we can calculate how much water it holds:

Dimensions shown in mm, convert: **1m = 1 000 mm.**
1m³ = 1 000 liters

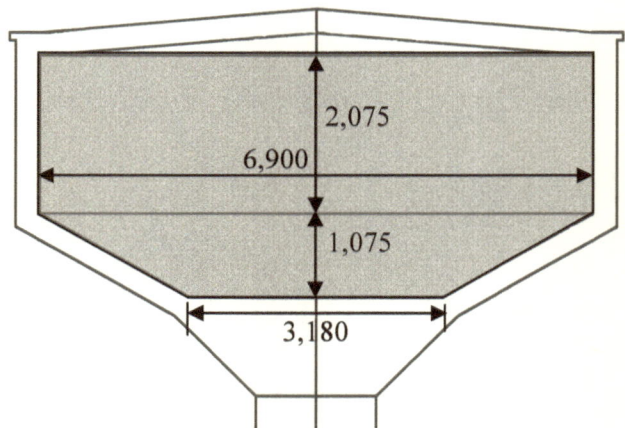

Do you see the two shapes we have to calculate? A *cylinder* and a *frustum of cone*.

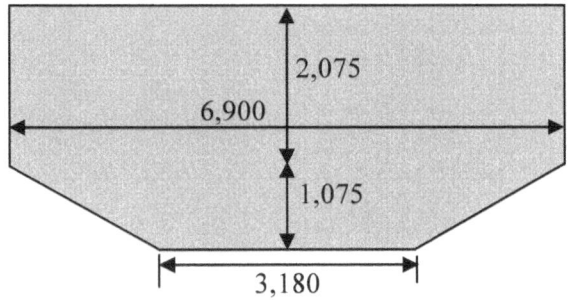

Volume of Cylinder $= \pi\, r^2\, h$

$$= \pi \times (3.45m)^2 \times 2.075m$$

$$= \mathbf{77.59m^3}$$

Volume of Frustum $= \dfrac{\pi\, h}{3}\ (R^2 + R\,r + r^2)$

$= \dfrac{\pi \times 1.075}{3}\ [(3.45m)^2 + (3.45m \times 1.59m) + (1.59m)^2]$

$$= \mathbf{22.42m^3}$$

(If you forget the formula, just use the 3 cones within the frustum; that's all the formula does.)

Cylinder + Frustum of Cone:

$$77.59m^3 + 22.42m^3 = 100.01m^3$$

Cubic Meters to Liters:

$100.01m^3 \times 1\ 000\ litres/m^3 = \mathbf{100\ 010\ litres}$

Area for 3-D Circular Shapes

Area of Sphere

Now what would be the chance that the formula for surface area around a *3-D sphere* will be *almost identical to* the formula for area of a *2-D circle*?

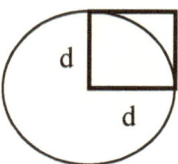

The only difference is that *radius becomes diameter*:

Area of Sphere = π d²

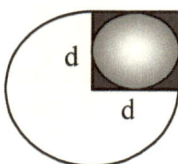

See the logic here: the outer circle was literally the 2-D circle, but with this 3-D object the sphere is inside the square, the *outer circle merely being equated to area of sphere*.

Area of Cylinder

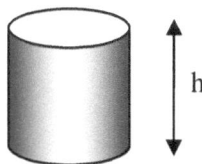

The logic here is that we basically *cut and roll open* the cylinder so that it becomes a *rectangle*:

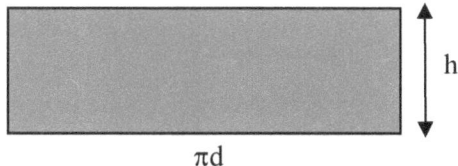

πd

The length of the rectangle will be *perimeter of circle πd*; thus:

Area of Cylinder = πdh

Area of Cone

However intimidating this shape looks, it's the very *same logic* as for area of a *cylinder*.

All we need here is to configure an *average cylinder*:

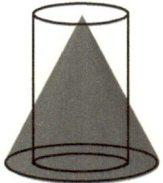

What is the average between diameter and zero diameter?
Radius, of course, meaning average diameter midway up.

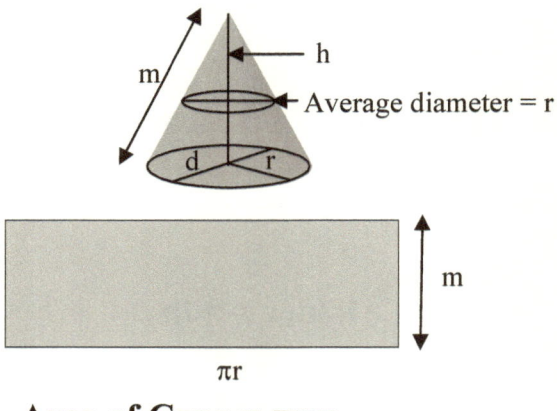

Area of Cone = πrm

Slope distance (m) may be misunderstood here as contradicting the golden rules for area calculation, but not really, the logic being that we want to create a rectangle with a perpendicular height equivalent to the *slope distance*. To calculate the slope distance is also just a logical procedure; we will deal with that in "Theorem of Pythagoras."

$$m = \sqrt{r^2 + h^2}$$

BASIC GEOMETRY

Let's familiarize ourselves with some of the fundamental concepts in geometry.

This is a *90° angle*, also named right angle. These lines are perpendicular to each other and are generally indicated by a little square in the corner of the angle.

If one of the corners of a triangle is a 90° angle, it is called a *right-angled triangle*.

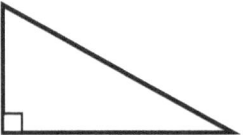

Angles on a straight *line equal 180°*.

¾ *turn* = 270°

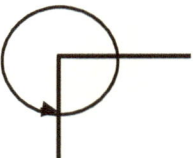

Angles on *full circle* = 360°

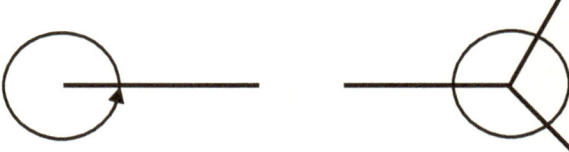

Parallel lines are lines that are spaced equally apart at any point on the line, i.e., however far we extend these lines they will never meet, generally indicated by arrows on the lines.

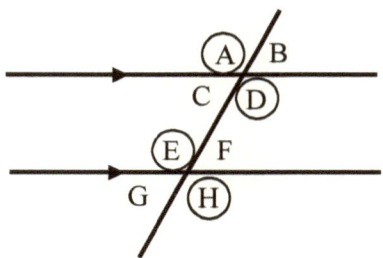

In the above the following are some of the *relationships* between the angles:

A = D (Opposite Angles)
A = E (Corresponding Angles)
D = E (Alternate Angles)

See how all the circled and the uncircled symbols have the same angles respectively.

How many degrees are the sum of the internal angles in a triangle?

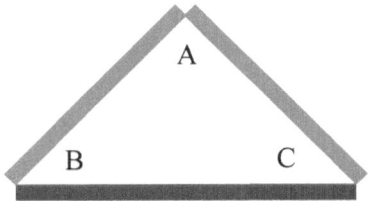

Let's see what happens if we press this triangle flat.

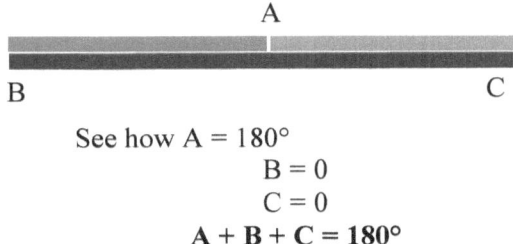

See how A = 180°
B = 0
C = 0
A + B + C = 180°

Example:

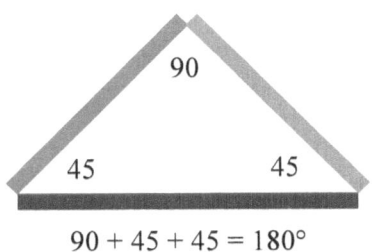

90 + 45 + 45 = 180°

This is true for any internal angles in a triangle:
The sum of the internal angles of a triangle = 180°

How many degrees are the sum of the internal angles in a 4-sided shape (square/rectangle/parallelogram/trapezium)?

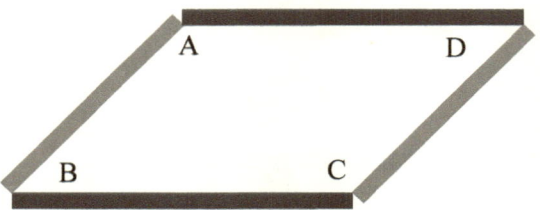

Let's press this 4-sided shape flat.

The lines can be presented in any order and stretched to any length, which will yield the same result.

See how we have 2 angles of 180° and 2 of 0°.

A + B + C + D = 360°

How many degrees are the sum of the internal angles in a 5-sided shape (regular or irregular pentagon)?

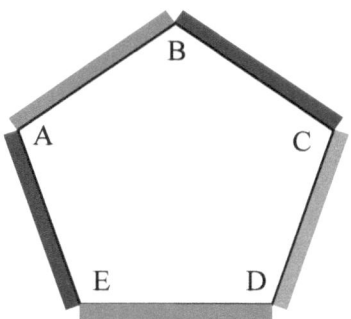

Let's press this pentagon flat.

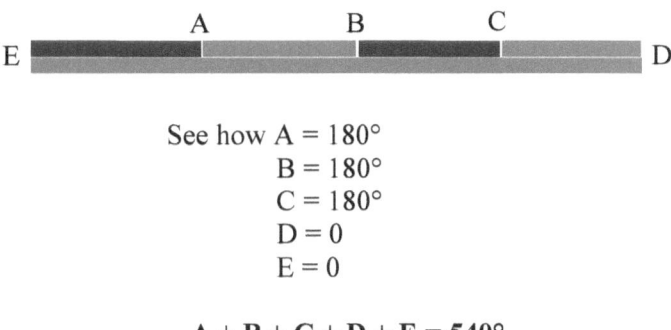

See how A = 180°
B = 180°
C = 180°
D = 0
E = 0

A + B + C + D + E = 540°

Do you *see* the pattern? Never mind how many sides a shape have, 2 angles can be equated to 0, and the rest will be 180° each.

CONFIGURATION OF 90° ANGLES

Whether it is a flower box, garage, shopping complex, or multistory building, no one wants to create *structures that resemble parallelograms*, if the *intention* was to create a square structure (square in this context have nothing to do with all sides being equal but merely that all lines are *perpendicular* [90°] to each other).

Just imagine the *disaster* if we built the complete structure of a building and then start laying tiles, and this is what we see!

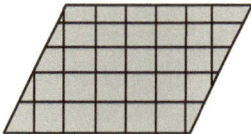

The illustration is a little bit exaggerated just to get the point across; in practice it would be almost invisible to the naked eye, until we start laying tiles that is. There are a number of ways to ensure that this never happens to us, ever.

Builders' Square

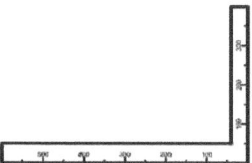

This tool is usually used for marking on boards and squaring of *little box-shaped objects*. For larger projects this is the least effective tool. Just imagine marking out 2 perpendicular lines *600mm* and *400mm* long, then trying to extend those lines to 10 meters both sides, the margin of error would definitely be too great for comfort.

Cross Measuring

This is one of the ways that a box-shaped object can be checked for squareness, but it is not recommended for initial setting out or marking out because we would have to create a box *more or less* in the right position first, then cross measure to set it straight. See how we are using the *diagonal lines of a rectangle*, which are always equal?

Isosceles Triangle

This is one of the most effective ways to create perpendicular lines for small- to medium-sized projects although not many people use it or are even aware of this option.

If we can remember the first time we used our mathematical sets in primary school.

Do you remember this?

Using a *compass* to mark the same distance both sides of a point on a line, then increasing the radius from these new points to create an arc intersection above/below the line. The resultant line from this intersection would then be perpendicular to the center point on original line. In practice a *tape measure* would do the same job that the compass did.

Sure you do, but sadly this, to most people, stayed a mathematical set exercise from way back.

Can you see what that exercise was all about?

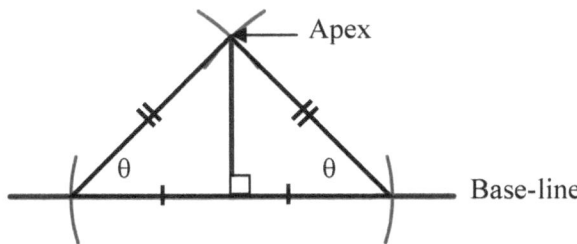

Creating an *isosceles triangle* and then halving it. An isosceles triangle has two legs of equal length and two equal angles. The line that runs from the apex will be perpendicular to the baseline.

Let's see how that simple concept can assist with this little do-it-yourself job at home:

We want to create a paved walkway 90° to our front door. How will we ensure that we don't have a skewed walkway?

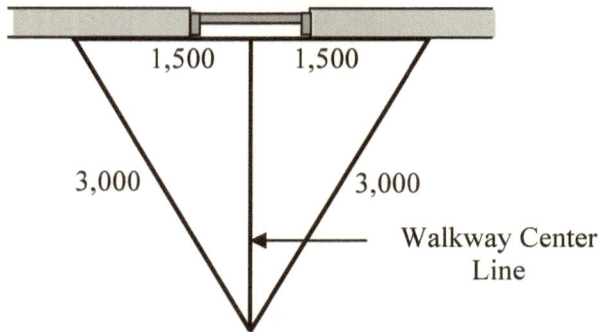

Do you see the isosceles triangle we created (as seen from above in this plan view)? Any dimensions will do, as long as we create an isosceles triangle. The center line is 90° from the center of door

See the end result, great job! And done right first time!

3:4:5 (A Derivation from Pythagoras)

This is a well-known and widely used ratio on small- to medium-sized projects in construction.

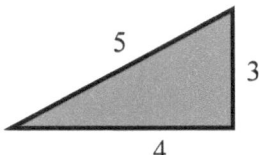

Let's see how we could increase the numbers without invalidating the ratio:

3 : 4 : 5

6 : 8 : 10   (each number multiplied by 2)

30 : 40 : 50   (each number multiplied by 10)

1.5 : 2 : 2.5   (each number divided by 2)

Although this ratio offers some flexibility in the magnitude of the numbers, it is limiting because of the constant angle.

Now we go to the ultimate procedure for finding and working with 90° angles, by the logic of the theorem of Pythagoras.

THEOREM OF PYTHAGORAS

Understanding the Theorem

This is the most useful theorem we will ever learn because there are no limitations. We can simply use the sides of any right-angled triangle and configure the diagonal line.

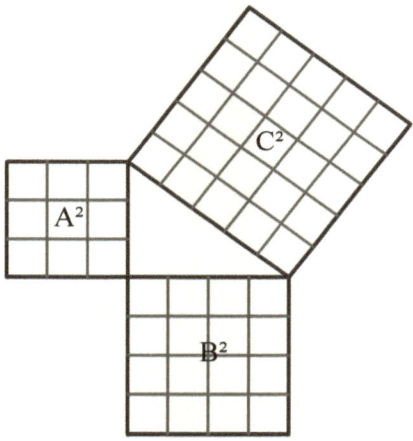

In the illustration of this right-angled triangle surrounded by a square block on each side, count the squares in block A^2 *and* B^2, and then add them together, $9 + 16 = 25$. See how the diagonal block C^2 have the same amount of squares, 25.

Believe it or not, that is *in short* the theorem of Pythagoras, that the sum of the square blocks on the perpendicular sides of a right-angled triangle equals the square block on the diagonal side.

$$A^2 + B^2 = C^2$$

Can you see the 3-D algebra in this formula?

The basic logic of calculating with Pythagoras would be:

- We would be given a triangle with *2 known* and *1 unknown* value.

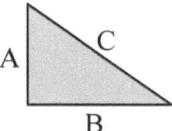

- Then we would create *square blocks* around all sides. The formula will give us the value of the unknown block (we may have to manipulate the formula to move the unknown to one side).
  $$A^2 + B^2 = C^2, \quad A^2 = C^2 - B^2, \quad B^2 = C^2 - A^2$$

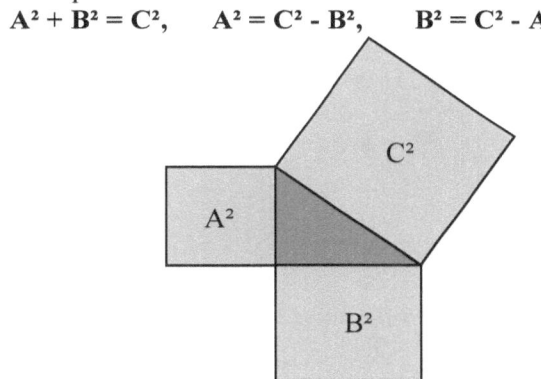

- Then we configure the *square root* of the unknown.

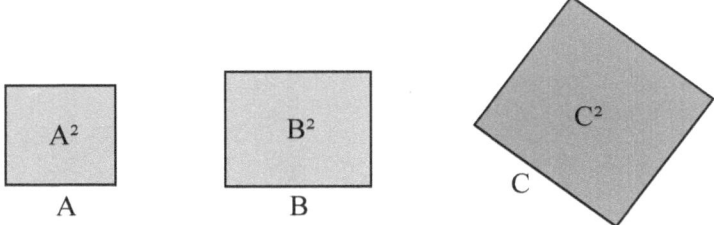

Let's take the same triangle where we were counting the blocks on, to see how the formula is applied:

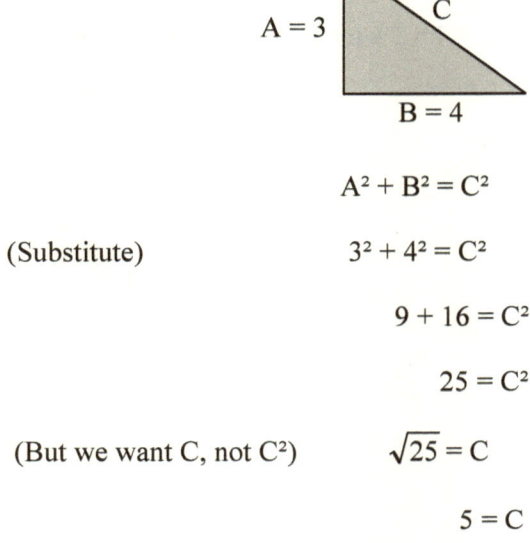

$$A^2 + B^2 = C^2$$

(Substitute) $\quad\quad\quad\quad 3^2 + 4^2 = C^2$

$$9 + 16 = C^2$$

$$25 = C^2$$

(But we want C, not C^2) $\quad \sqrt{25} = C$

$$5 = C$$

See how we proved *3:4:5*, but the great news is it works the same way for any numbers. Try substituting your own numbers to see how this works. Use a calculator. $A^2 + B^2 =$, then get the square root ($\sqrt{}$) of the result.

Manipulating this formula is quite simple, nothing to it really. Inverse of addition is subtraction.

$$A^2 + B^2 = C^2$$

$$A^2 \quad\quad = C^2 - B^2$$

Or $\quad\quad\quad\quad$ --------------------

$$A^2 + B^2 = C^2$$
$$B^2 = C^2 - A^2$$

Configuration of Circle with Pythagoras

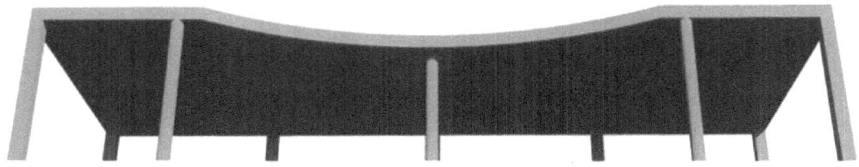

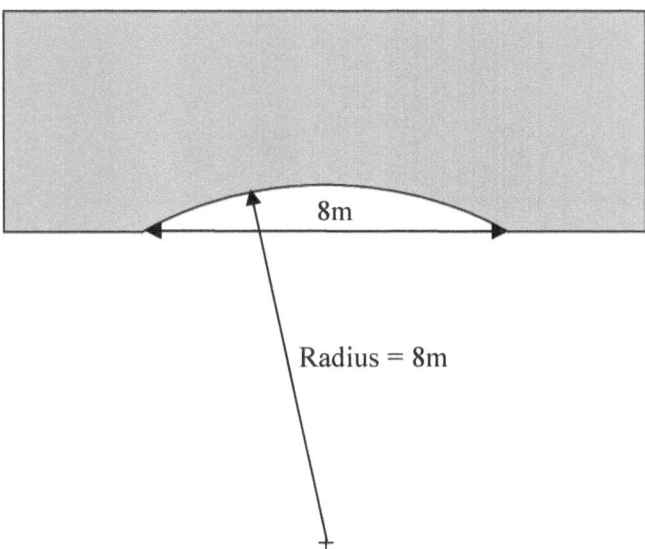

This slab needs to be built as the *1st floor* to an office block under construction and consists of a circle segment where the *center point* is not on the slab but outside in midair, approximately 4 meters off the ground. We have constructed the deck where the slab will be cast on. Now how on earth will we *mark* this circle segment? However unbelievable it may sound, the answer lies with Pythagoras.

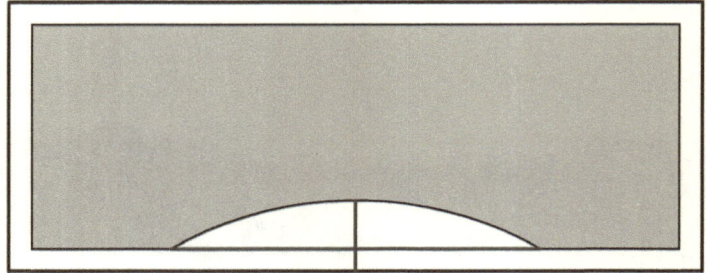

Now, just to get the general idea, shown above, the outer rectangle is the *deck* where the slab will be cast on. The lines within the segment represent the logical *reference lines* for the circle segment under the current conditions.

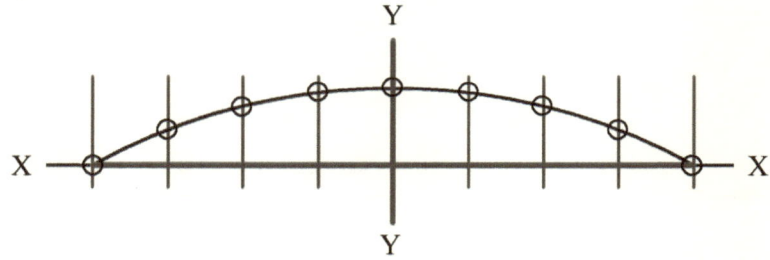

Now, firstly we need to draw a *grid* as shown, to create reference *points* on the circle; we've used a 1-meter grid here, but in practice we could use an even smaller grid, like 250mm.

Then, common sense should tell us that we need to find the dimensions from these intersections with the curve to the X line so that we can mark these points from on top of the deck.

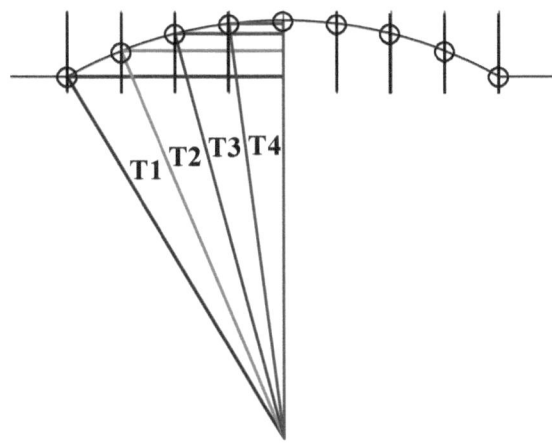

Can you see the logic? How every *triangle has 2 known sides and 1 unknown, radius* being the *hypotenuse* or C side, 8 meters, and the grid markings being the other side, 1m, 2m, 3m, and 4m respectively? Obviously we will do only 1 triangle at a time.

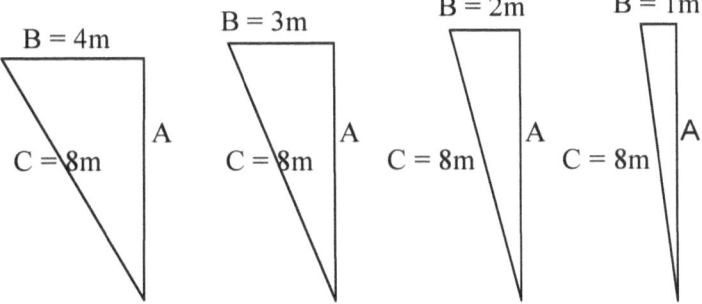

See how A is the unknown in all the triangles.

The calculations are as follows:

$$A^2 + B^2 = C^2$$

$$A^2 = C^2 - B^2$$

$$A = \sqrt{C^2 - B^2}$$

T1 $\quad A = \sqrt{8^2 - 4^2}$
$\qquad\quad = 6.928$

T2 $\quad A = \sqrt{8^2 - 3^2}$
$\qquad\quad = 7.416$

T3 $\quad A = \sqrt{8^2 - 2^2}$
$\qquad\quad = 7.746$

T4 $\quad A = \sqrt{8^2 - 1^2}$
$\qquad\quad = 7.937$

With that we have configured the circle coordinates *relative to the circle center point*. Now we need to relate that dimensions to our X line. The circle center to X line = 6.928, can you see that? The 1st triangle, T1, was configured on the X line. Now we need to subtract this from all the configured dimensions.

T1 $\qquad$ $6.928 - 6.928 = \mathbf{0}$

T2 $\qquad$ $7.416 - 6.928 = \mathbf{0.488}$

T3 $\qquad$ $7.746 - 6.928 = \mathbf{0.818}$

T4 $\qquad$ $7.937 - 6.928 = \mathbf{1.009}$

$\qquad\qquad$ $8.000 - 6.928 = \mathbf{1.072}$

See how the last dimension is not a triangle but the center line, equal to radius 8m.

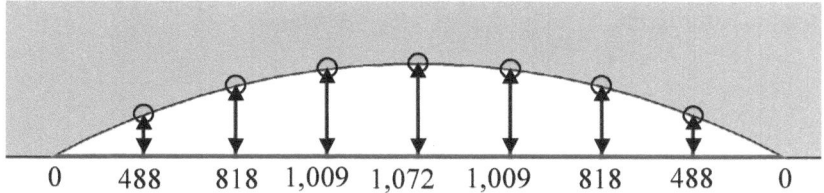

| 0 | 488 | 818 | 1,009 | 1,072 | 1,009 | 818 | 488 | 0 |

More than any other field in real life, *construction is the playground of mathematics*, where the only limitation to solving everyday complexities is our level of comprehension of mathematical principles and how well we can adapt to *seeing* the mathematics.

That concludes our in-depth look at theorem of Pythagoras, but it will definitely not be the last time we use the services of this loyal servant of mathematics, always just a few blocks away.

WORKING WITH OTHER COMMON ANGLES

45° Angle

This is probably the most commonly used angle after 90°. If we want to configure a 45° angle, we would simply measure the same distance onto two perpendicular lines.

The diagonal line closing the triangle will be 45°.

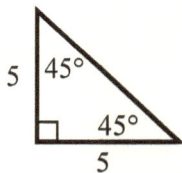

If we want to know what the diagonal dimension is, we simply use Pythagoras:

$$A^2 + B^2 = C^2$$

$$5^2 + 5^2 = C^2$$

$$25 + 25 = C^2$$

$$50 = C^2$$

$$\sqrt{50} = C$$

$$7.071 = C$$

But what if only the diagonal is known?

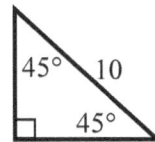

The logic here is that A = B

So we could say A + B = 2A
Or   A + B = 2B

$$2A^2 = C^2$$

$$A^2 = \frac{C^2}{2}$$

$$A^2 = \frac{10^2}{2}$$

$$A^2 = \frac{100}{2}$$

$$A = \sqrt{50}$$

$$A = 7.071$$

That's it for 45° angles.

30° and 60° Angles

These are probably the most commonly used angles after 45°. If we want to configure a 30°/60° angle onto two perpendicular lines, we need to draw a horizontal line of any dimension, rather longer than shorter; then start by measuring a distance onto the vertical:

Then we measure twice that distance from the top of vertical line diagonally down onto the horizontal:

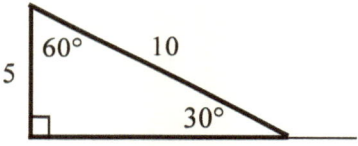

Result: 30°/60° triangle

Aha! You guessed right, we need to see the logic behind this.

Right, let's make a mirror image of this triangle below the horizontal line.

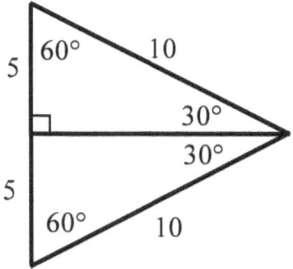

Aha! Do you *see* the logic?

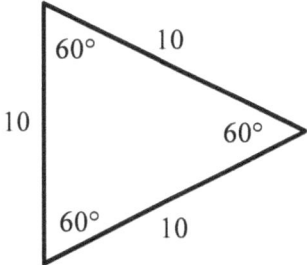

Do you see how all sides are equal in this *equilateral triangle*? Do you see how a 30°/60° triangle is half of an equilateral triangle? Great stuff!

Let's now move on to the exciting world of trigonometry.

BASIC TRIGONOMETRY

The Ratios

Trigonometry is the study of the relationship between the sides and the angles in a right-angled triangle. We need to familiarize ourselves with the diagram and understand the *terminology* related to it.

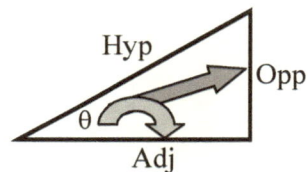

Angle θ - (Theta)
Hyp - (Hypotenuse)
Opp - (Opposite)
Adj - (Adjacent)

Hypotenuse is always the diagonal side, also the longest side.
Opposite is always the opposite side from θ.
Adjacent is always the adjacent side to θ.

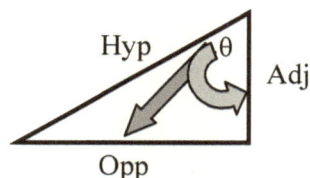

See how Opp and Adj is always relative to θ.

Now, before we continue, we need to clearly *see* the following story in our minds; let's pay careful attention to this:

It's about road rage.

The door on the driver's side of the pickup truck is flung open and the driver is *standing* on the ledge, fist in the air, showing *signs* of aggravation toward a passing car. Let's call this driver the sign man.

Next to him in the truck, *sitting in bent position*, is his *codriver*, also showing signs at the passing car. We will call him the *cosign man.*

At the back of the pickup truck there is another guy, apparently undisturbed by the commotion going on in the front, dressed only in shorts, he is relaxing, *legs stretched out*, basking in the sun, having a *tan*. This guy we will call the *tan man.*

Now, what have we learned from this story? Believe it or not, we have just learned how to instantly recognize the ratios in any right-angled triangle. Look at the pictures below and see if you recognize the *sign man, cosign man*, and *tan man* from their postures in the story.

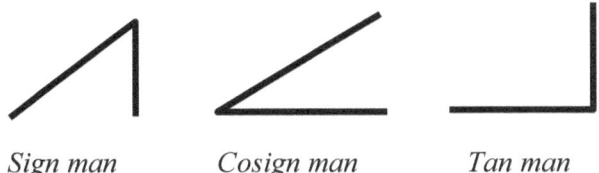

Sign man   *Cosign man*   *Tan man*

Now, let's look at our triangle:

Look for the *sign man*, *cosign man*, and *tan man* in this triangle. Great stuff! I knew you would see them instantly.

Now, how do the ratios work?

$$\boxed{\mathbf{Sin\,\theta \;=\; \dfrac{Opp}{Hyp}}}$$

See, the *sign man* is actually the sine of theta (Sin θ). Then just think of the hypotenuse as a *hippopotamus*, the heavy one, always *sinking* to the bottom.

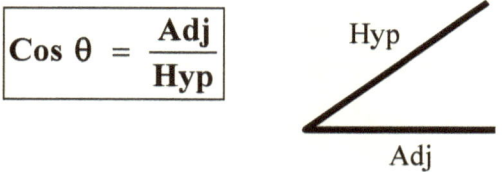

$$\boxed{\mathbf{Cos\,\theta \;=\; \dfrac{Adj}{Hyp}}}$$

The *cosign man* is the cosine of theta (Cos θ). See once again how the *hippo* sinks to the bottom.

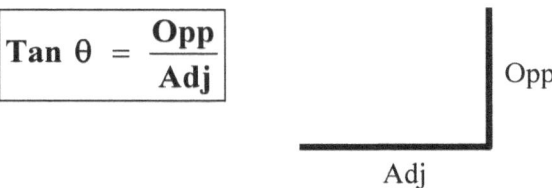

The *tan man* is tangent of theta (Tan θ). Now, just say *Opp* aloud; doesn't that sound a lot like *up*? Right, just remember that Opp always goes *up* to the top.

Great guns! Who said we can't *see* trigonometry!

Now, when can we use trigonometry? If we have at least 2 known sides

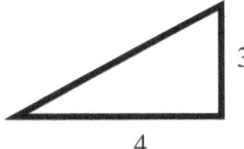

or a side and an angle.

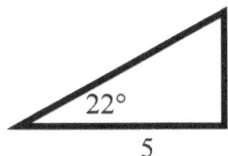

In both the above illustrations it is possible to solve any unknown side or angle.

Finding an Angle

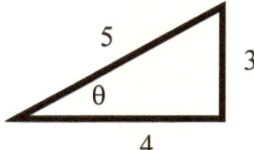

As we can see in this triangle all the sides are known, so we could use any of the ratios to determine the angle. Let's use the *sign man*:

$$\text{Sin}\,\theta = \frac{\text{Opp}}{\text{Hyp}}$$

$$= \frac{3}{5}$$

$$= 0.6$$

See how 0.6 is the sine of the angle θ; now to find the angle we need to find the *arcsine* of 0.6 ($\text{Sin}^{-1}\,0.6$), also called inverse sine.

So:

$$\text{Sin}^{-1}\,0.6 = 36.87°$$

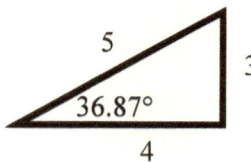

Or we could have gone straight to the arc ratio from the start, thus:

$$\operatorname{Sin}^{-1}\left(\frac{\text{Opp}}{\text{Hyp}}\right) = \theta$$

$$\operatorname{Sin}^{-1}\left(\frac{3}{5}\right) = 36.87°$$

$$\operatorname{Cos}^{-1}\left(\frac{\text{Adj}}{\text{Hyp}}\right) = \theta$$

$$\operatorname{Cos}^{-1}\left(\frac{4}{5}\right) = 36.87°$$

$$\operatorname{Tan}^{-1}\left(\frac{\text{Opp}}{\text{Adj}}\right) = \theta$$

$$\operatorname{Tan}^{-1}\left(\frac{3}{4}\right) = 36.87°$$

Applying Trigonometry on Roof Trusses

Let's see some examples of how trigonometry is utilized in the real world:

We have to build roof trusses for a double garage, so we decide to build the trusses on site to save costs. On the layout drawings the span is 6 meters, *roof pitch 22°. Height* is unknown, needs to be configured.

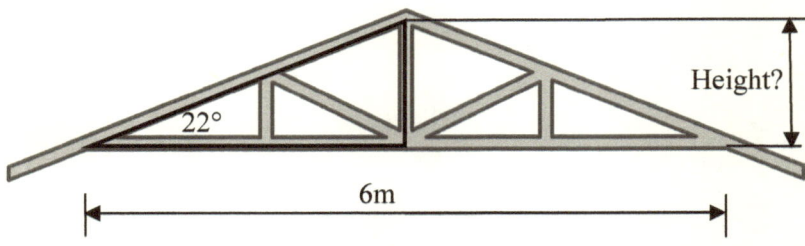

Let's draw our triangle to see this more clearly. Note how our triangle is only half the span.

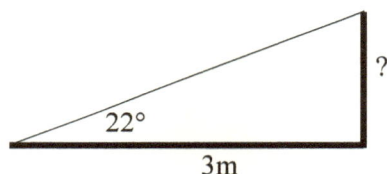

Now which ratio are we going to use? Aha! You saw the *tan man* didn't you?

$$Tan\ \theta\ =\ \frac{Opp}{Adj}$$

$$Tan\ \theta\ \times Adj\ = Opp$$

$$Tan\ 22\ \times 3m\ = 1.212m$$

Bevel Cut

The illustration shows the *bevel cut* we need to make so that the two rafters can butt joint neatly together. Same roof, 22° pitch. We want to use a builders' square for this. How?

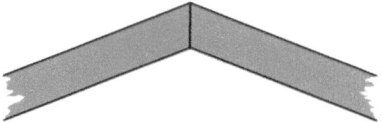

Builders' squares have measurement markings on them, same as a ruler but both sides and on both arms. If we can figure out which dimensions to use to create a 22° angle, then Bob's your uncle.

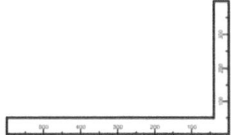

Let's draw the triangle. Then we have to *assume 1 measurement*; let's try 200mm vertically, and then we need to configure the horizontal measurement.

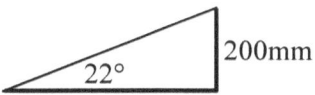

200mm

$$\text{Tan } \theta = \frac{\text{Opp}}{\text{Adj}} \quad \text{(Eish! Adj needs to be above the line and alone)}$$

$$\text{Adj} = \frac{\text{Opp}}{\text{Tan } \theta} \quad \text{(See how } \textit{Adj} \text{ and } \textit{tan } \theta \text{ traded places)}$$

$$\text{Adj} = \frac{200}{\text{Tan } 22} = 495\text{mm}$$

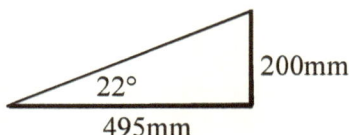

Right, got it; now let's *mark* these dimensions on the builders' square with a marker:

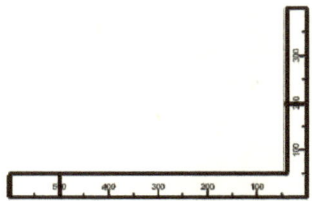

Then we mark the rafter.

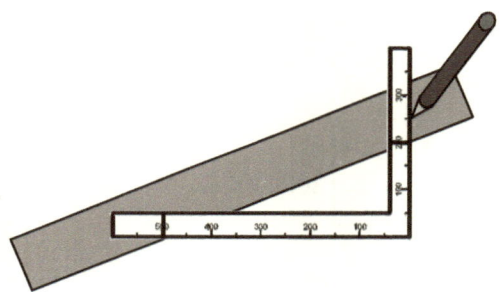

See how we turned an ordinary builders' square into an amazing *multiangle tool* with trigonometry!

Logical Thinking with Frustum

In mathematics there are always alternative ways of overcoming problems; if, for example, we *forgot* a formula, that shouldn't stop us from solving a problem with logical thinking.

Let's take volume for a frustum of cone for instance:

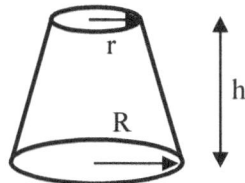

What is a frustum of a cone? A *portion* of a cone, of course; thus, if we could configure what the full cone's volume was and then deduct the little cone on top of the frustum, that should give us volume of frustum, right?

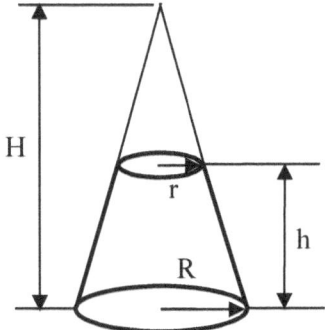

Now how will we know what the full cone's height is? That's simple really; we have R (base radius), now all we need is an *angle*, then we use one of the trig ratios to get height.

If we deduct r from R, that should give us a *little triangle* as shown, where Opp and Adj is known; thus θ is configurable.

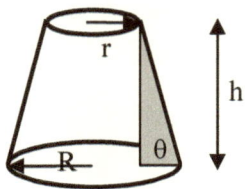

$$\theta = \text{Tan}^{-1}\left(\frac{\text{Opp}}{\text{Adj}}\right)$$

Then that angle can be used to configure the H (height) of the cone.

$$\text{Tan}\ \theta \times R = H$$

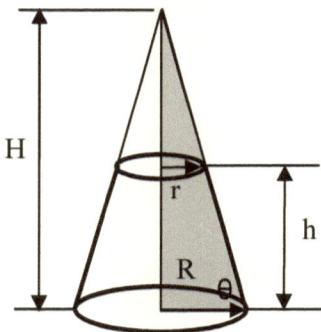

$$\text{Volume of frustum} = \frac{\pi R^2 H}{3} - \frac{\pi r^2 (H-h)}{3}$$

Do you *see* how we didn't need the volume of frustum formula here?

Logical Thinking with Segments

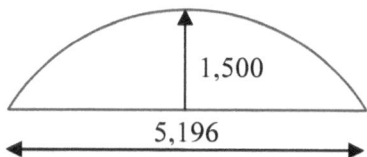

We want to calculate area of this segment. To be honest there is quite an intricate formula for this, which we will most probably forget very quickly, so let's rather focus on the logic.

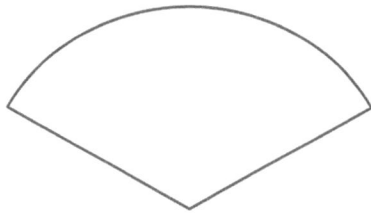

We need to complete the sector and configure radius first; otherwise area will be unconfigurable.

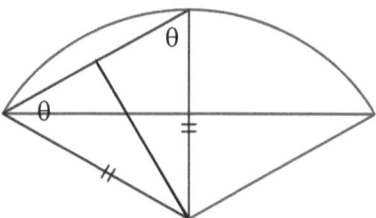

Do you see the isosceles triangle (two equal sides)? Meaning also two equal angles, which are unknown, and so are the legs.

Can you see how that triangle is the key to finding radius if we can find the unknowns?

So let's focus our attention on another little triangle first; do you see how 2 dimensions in this triangle are known? So we could easily configure the 3rd and the angle.

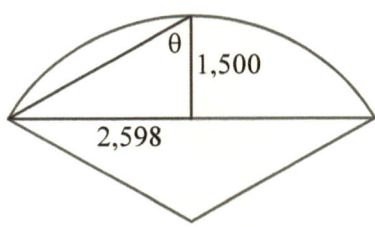

Let's do the side first with Pythagoras:

$$A^2 + B^2 = C^2$$

$$\sqrt{2,598^2 + 1,500^2} = \sqrt{C^2}$$

$$C = 3,000$$

Now the angle:

$$\theta = \tan^{-1}\left(\frac{Opp}{Adj}\right)$$

$$\theta = \tan^{-1}\left(\frac{2,598}{1,500}\right)$$

$$\theta = 60°$$

Aha! Do you see what I see?

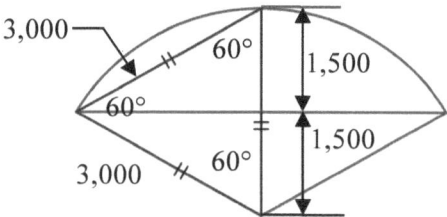

Of course, if 2 angles in a triangle = 60°, then so are the 3rd, i.e., our triangle is actually an equilateral triangle (all sides equal). That means even the sides are the same! So radius = 3,000. See how half the side is 1,500 (as per our given dimension).

See how easy this calculation is now:

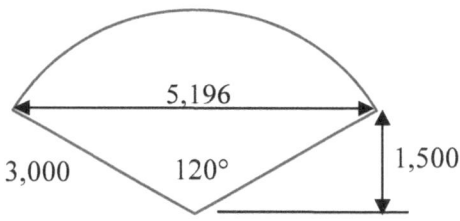

Area of sector: $\dfrac{120}{360} \pi \, 3^2 = 9.425$

Area of triangle: $0.5 \times 5.196 \times 1.5 = 3.897$

Area of segment:

Area of sec tor − Area of triangle

$= 9.425 - 3.897$

$= 5.528$

By some strange coincidence, we found an equilateral triangle in our calculation; why?

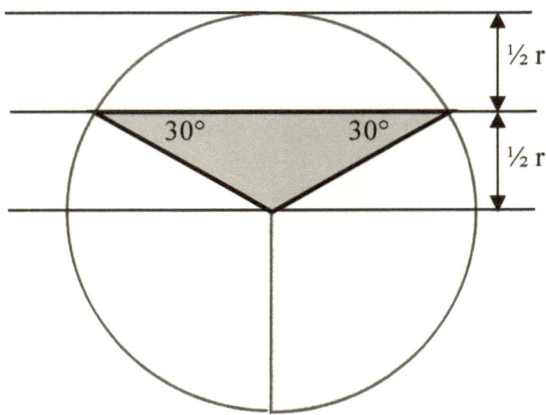

Aha! See the creation of the segment, how our segment is half radius? That automatically creates an *isosceles triangle* with 2 angles being 30°!

So what if this was not the case? Like if the segment height was not half radius?

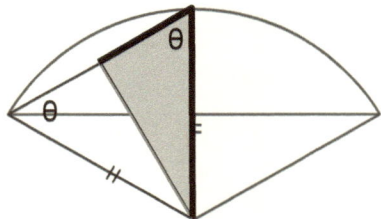

We would've needed to calculate radius by configuration of the shaded triangle as shown. Do you see the upside-down cosine ratio (cosign man)? Just plain old logic, isn't it?

Plotting Points on a Circle

Okay, we've seen how it works; now let's take trigonometry a little bit
further into the playgrounds of construction.

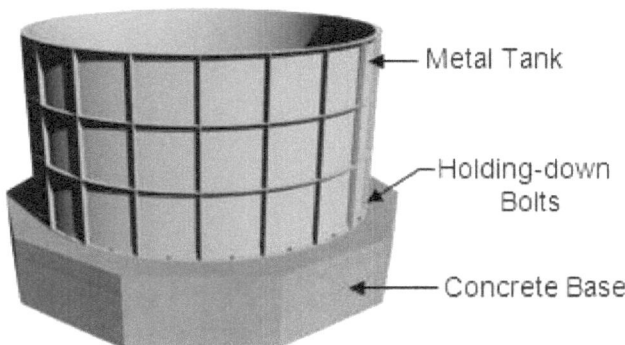

Metal Tank

Holding-down
Bolts

Concrete Base

What we are looking at is how a huge metal tank needs to sit on a massive
concrete base as shown. The tank will be secured to the base by *holding
down bolts* that need to be positioned and cast into the concrete while
constructing the base. The tank is still being manufactured elsewhere, so all
we have to work with is the *engineer's specifications*.

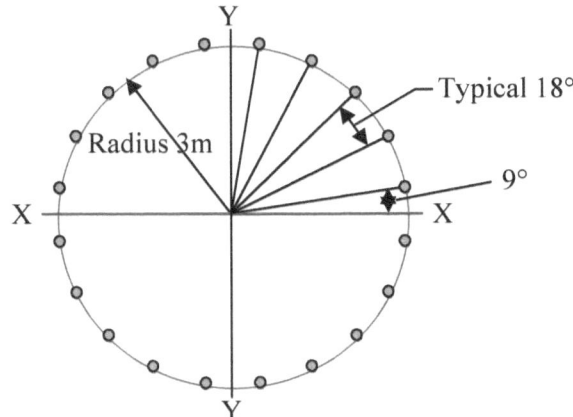

Typical 18°

Radius 3m

9°

We have to construct the base and ensure that the bolts are in the right position to accommodate the tank. How?

The only way to accurately *plot* the bolt points would be to relate each point to the X and Y center lines.

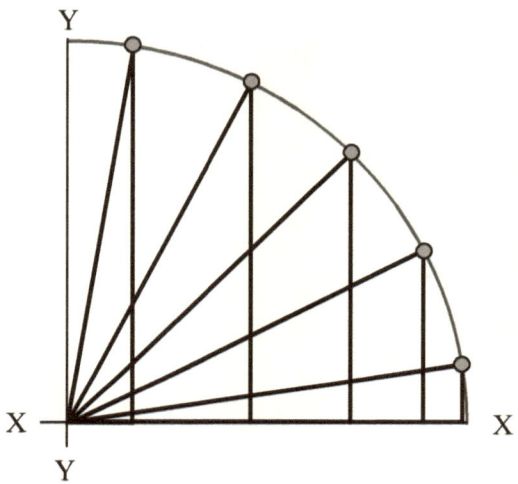

See the triangles? How each triangle has 1 known side, *radius* of circle being *equal* to *hypotenuse* in each triangle, 3 meters? Angle of θ is also configurable for each triangle, right?

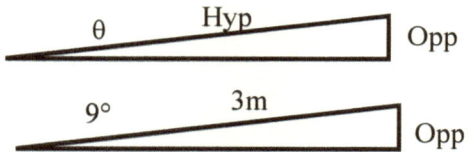

See the sign man?

$$\text{Sin } \theta = \frac{\text{Opp}}{\text{Hyp}}$$

$$\text{Sin } \theta \times \text{Hyp} = \text{Opp}$$

Angle: **9°**

Sin 9 x 3m = **0.469m**

_____

Angle: 9° + 18° = **27°**

Sin 27 x 3m = **1.362m**

_____

Angle: 27° + 18° = **45°**

Sin 45 x 3m = **2.121m**

_____

Angle: 45° + 18° = **63°**

Sin 63 x 3m = **2.673m**

_____

Angle: 63° + 18° = **81°**

Sin 81 x 3m = **2.963m**

_____

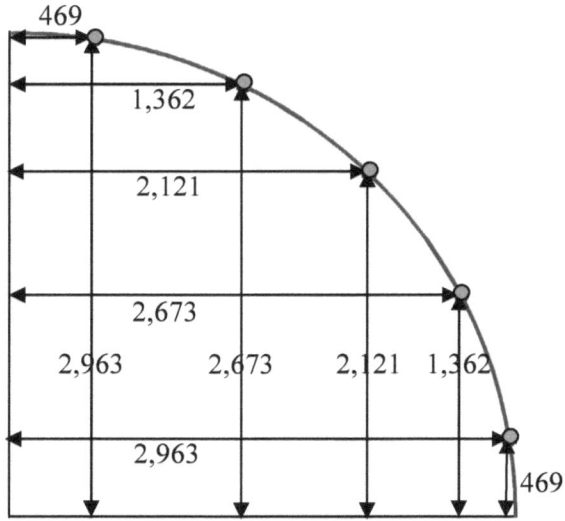

Can you see how we configured all the dimensions in the quarter circle, the counterclockwise direction being a mirror image of the clockwise?

Then, the great news is that we have actually configured the whole circle, because each of the other quadrants will be a mirror image of this one!

Do you *see* the mathematics?

THE COSINE RULE

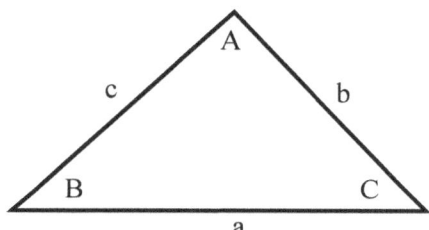

When it comes to working with nonright-angled triangles, the cosine rule is probably the most useful rule we will ever learn. In the above illustration the sequence of the symbols doesn't matter at all, as long as we remember:

Side a is always opposite Angle A.
Side b is always opposite Angle B.
Side c is always opposite Angle C.

Cosine Rule:

$$a^2 = b^2 + c^2 - 2bcCosA$$

$$b^2 = a^2 + c^2 - 2acCosB$$

$$c^2 = a^2 + b^2 - 2abCosC$$

This is the conventional way that this formula was learned and taught.

The Logic of the Cosine Rule

Unfortunately, for most of us this is probably the most useless rule we ever learned because if we may need it someday we will have forgotten the formula, but that might just change right here, today.

In this lesson we will discover what this formula actually means in layman's terms and where it originated from so that we may be empowered with the ability to solve problems by a logical procedure, instead of *substituting* numbers into a *formula*.

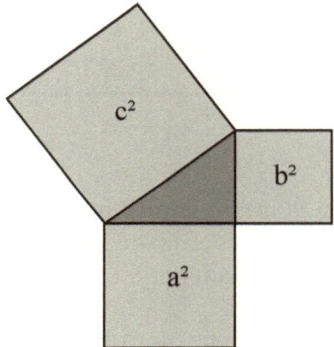

No, we are not repeating the lesson on theorem of Pythagoras, but it is indeed a relief that you have recognized this configuration (it means that you have been paying attention).

This theorem forms the basis of the cosine rule. Let's see how:

$$\mathbf{a^2 + b^2 = c^2}$$

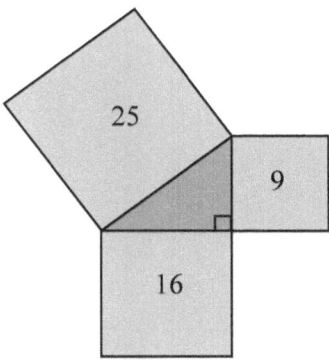

We should all be familiar with the 3:4:5 triangle by now, so let's use this as the example for our analysis.

$$a^2 + b^2 = c^2$$

Now let's take all the values to one side of this equation:

$$a^2 + b^2 - c^2 = 0$$

$$16 + 9 - 25 = 0$$

See how this equation is still true when we substitute numbers. So what does this mean? An answer of *zero* simply proves that the angle in question is *90°*.

Now, let's see what happens if we flip this equation to make it relative to the other angle, θ, instead of the 90° angle:

So, now we swap the equation around to make b = 9, the hypotenuse, c², as per the original equation. That will make the equation *relative to angle θ*.

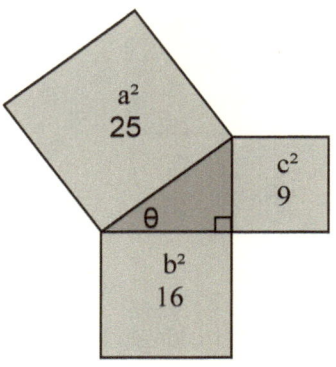

$$a^2 + b^2 - c^2 = ?$$

$$25 + 16 - 9 = 32$$

See that! That automatically means that we are not dealing with a 90° angle, but more importantly, this is the most significant number in the cosine rule, this *deviation* from Pythagoras.

Now, for practical reasons, which will become evident in a moment, we will introduce a completely new way of presenting the squares in this triangle.

Move the squares to new positions as shown.

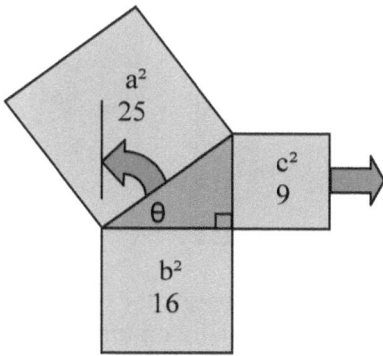

Resultant configuration

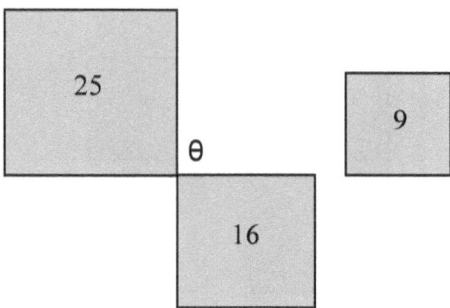

The reason for this new *alignment* of the squares is that we want to introduce two new rectangles into this configuration, thus:

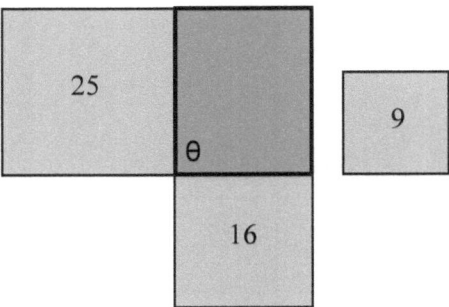

Okay, but that's only one rectangle; where's the other?
Good point, the two rectangles are actually on top of each other, so let's move the one rectangle off the other just a little bit for graphical purposes.

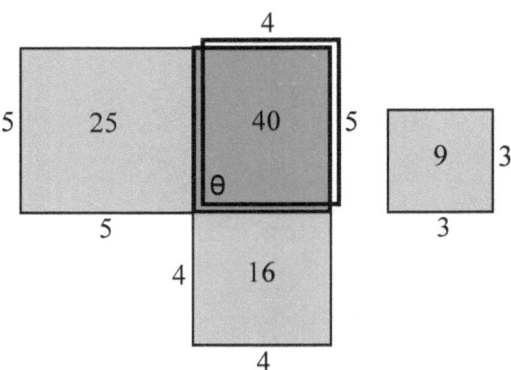

Good, that's more like it!
Those *two rectangles* are the 2nd most important factor in the cosine rule, and their values are determined by the adjacent squares. Can you see how these rectangles are *5 x 4 = 20 per rectangle*? So the *two* of them will *equal 40*, right? Great!

Now let's get to the cosine rule:

$$\text{Cos } \theta \ = \ \frac{\boxed{32}}{\square}$$

That should suffice in explaining the *cosine rule*. The cosine of the angle in question equals the deviation from Pythagoras divided by the two rectangles created from the two adjacent squares.

$$\text{Cos } \theta \ = \frac{32}{40} = 0.8$$

$$\theta \ = \text{Cos}^{-1} \ 0.8$$

$$= 36.87°$$

See how we configured this angle by the cosine rule? Obviously, because this is a right-angled-triangle we could have gone straight to the cosine trig ratio (*cosign man*), which would have yielded the same result.

$$\text{Cos } \theta \ = \ \frac{\text{Adj}}{\text{Hyp}} = \frac{4}{5} = 0.8$$

Let's create another triangle where this angle remains the same but we just change the 90° angle.

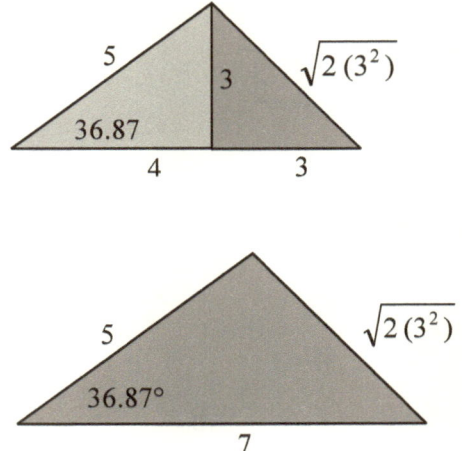

Right, see this new triangle? Notice how the values changed, so let's see what effect this will have on the procedure.

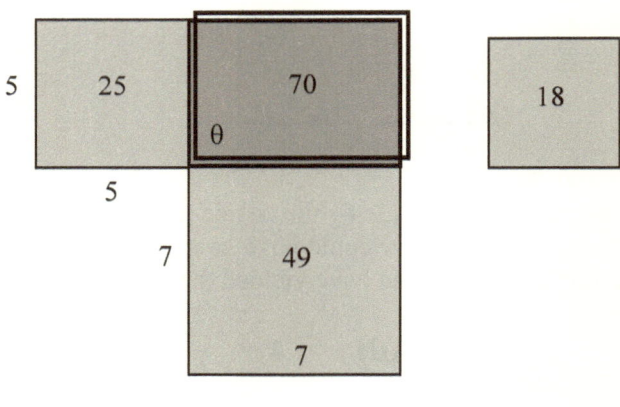

$$Cos\ \theta = \frac{25 + 49 - 18}{70} = 0.8$$

See! Same result! Like magic!

Now, in real life, shouldn't that be the way we should remember formulae, by the logic procedure behind it?

So for all intentive purposes, the cosine rule should be remembered like this:

$$\text{Cos } \theta = \frac{a^2 + b^2 - c^2}{2ab}$$

With this configuration of blocks:

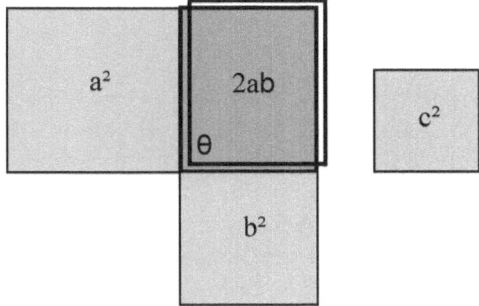

that could always be reverted back to the currently promoted formula by simple *manipulation*.

Can you *see* how this equation would be virtually impossible to explain by any means other than the blocks of algebra?

Area of Trapezium

Let's see how the cosine rule is applied in real life:

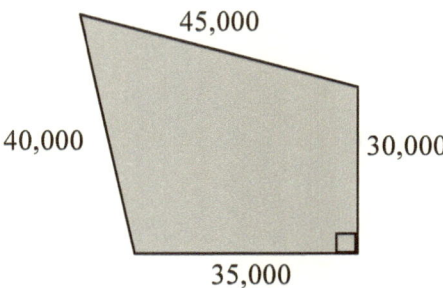

An estate agent is selling this piece of land; they say it is *approximately 1,000m²*. Now what does that mean? That they don't know their maths or possibly that it's probably less? Let's find out. This is not just a straightforward shape, so we have to give this a little thought. We have a 90° corner, so that should make the A triangle easy to configure. The trick here is to get the diagonal dimension so that we can use the cosine rule to configure another angle, then we will be able to calculate a perpendicular dimension to get area B.

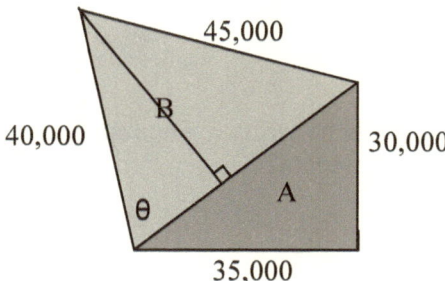

Right, let's do this.

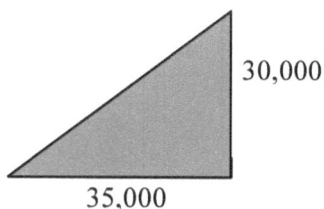

Pythagoras: $\qquad \sqrt{a^2 + b^2} = c$

$$\sqrt{35,000^2 + 30,000^2} = 32,404$$

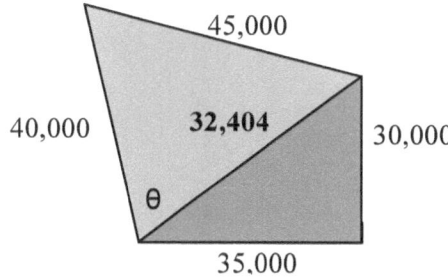

Right, now we just swing this triangle straight so that we can see this more clearly:

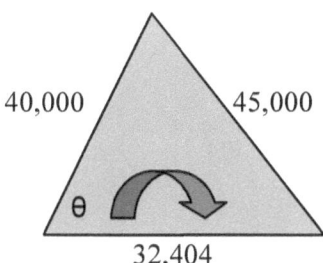

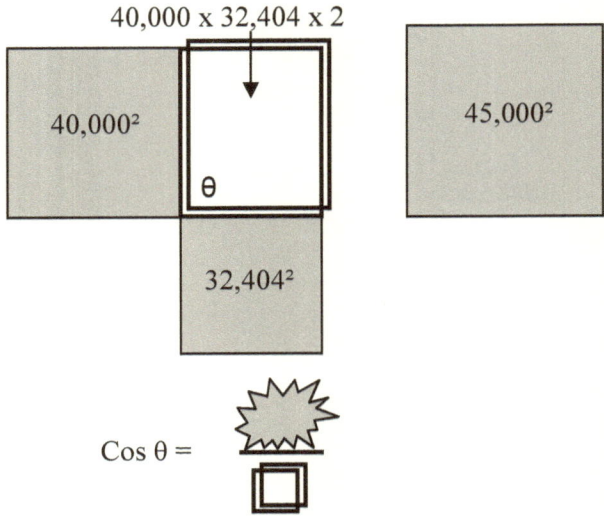

$$\text{Cos }\theta =$$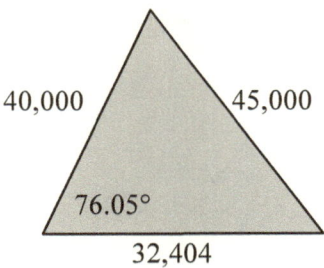

$$\text{Cos }\theta = \frac{40{,}000^2 + 32{,}404^2 - 45{,}000^2}{40{,}000 \times 32{,}404 \times 2} = 0.241$$

$$\theta = \text{Cos}^{-1}\, 0.241 = 76.05$$

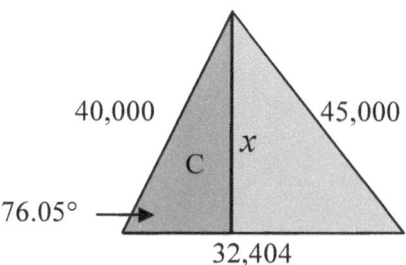

See how we now have another triangle (C), where *x* is configurable. Do you see the sign man?

$$\text{Sin } \theta = \frac{\text{Opp}}{\text{Hyp}}$$

$$\text{Sin } \theta \times \text{Hyp} = \text{Opp}$$

$$\text{Sin } 76.05 \times 40,000 = \text{Opp} = 38,820$$

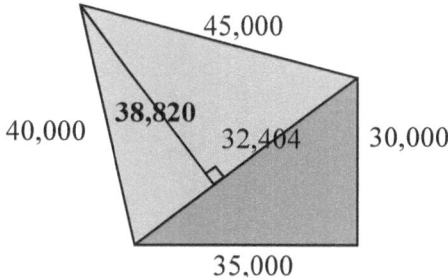

See how area is now configurable for both triangles!

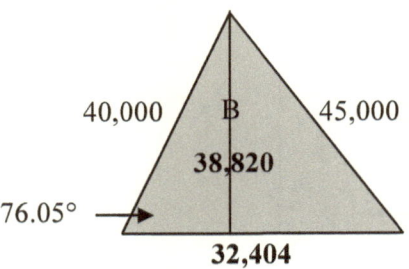

Area B = L x B x 0.5
  = 32. 404 x 38.820 x 0.5
  = 628.96 m²

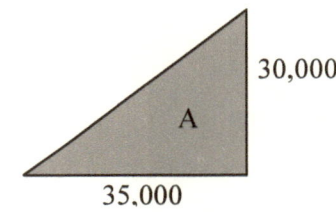

Area A = L x B x 0.5
  = 35 x 30 x 0.5
  = 525 m²

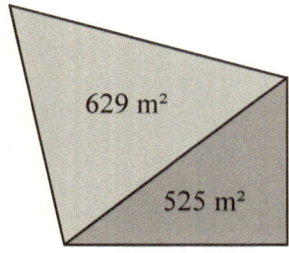

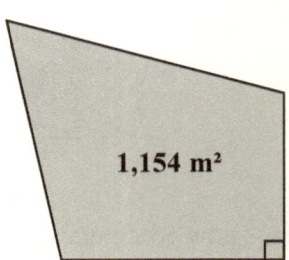

Aha! They can't calculate, we score! Benefit of the doubt, *154 m²* in our favor.

THE SINE RULE

The Logic of the Sine Rule

To better understand the logic of the sine rule, let's look at *right-angled triangles* and their *relationship* to a *circle*.

Study the illustration, see how *diagonal lines* in rectangle equals diameter of circle.

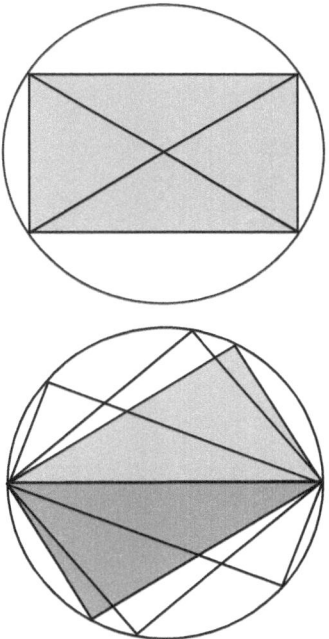

Can you see how a right-angled-triangle is created by splitting the rectangle into two parts? See how these triangles are all created on the diameter line of a circle and how all the *corners* are *touching* the *outer circle*? All right-angled triangles share this same relationship to circles.

Now let's get to the sine rule:

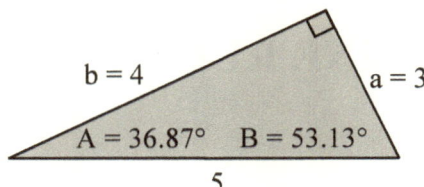

We've furnished all the values in this triangle to clearly see the inner workings of the sine rule.

$$\frac{a}{Sin\ A} = \frac{b}{Sin\ B} = \frac{c}{Sin\ C}$$

In our example we will use A and B:

$$\frac{a}{Sin\ A} = \frac{b}{Sin\ B}$$

$$\frac{3}{Sin\ 36.87} = \frac{4}{Sin\ 53.13}$$

Let's substitute values: $Sin\ 36.87 = \frac{3}{5}$, $Sin\ 53.13 = \frac{4}{5}$

$$\frac{3 \times 5}{3} = \frac{4 \times 5}{4}$$

Now, here we probably need to know the logic of dividing by a fraction before we continue.

Let's see what happens when we divide by a whole number:

$$3 \div 5 = \frac{3}{5} = \frac{3}{1} \times \frac{1}{5}$$

Do you see the pattern? When we divide by a whole number, we are actually multiplying by the inverse of that number, i.e.,

$\div 5$ becomes $\times \frac{1}{5}$

$$5 \text{ means } \frac{5}{1}, \text{ so } \div \frac{5}{1} \text{ becomes } \times \frac{1}{5}$$

So that pattern is the same when we divide by a fraction; we simply multiply by the inverse:

$$\frac{3}{1} \div \frac{3}{5} = \frac{3}{1} \times \frac{5}{3} = \frac{3 \times 5}{3}$$

Let's continue with our exploration into the inner workings of the sine rule:

$$\frac{3 \times 5}{3} = \frac{4 \times 5}{4}$$

See how Opp dimensions 3 and 4 are cancelled out of the equation? Leaving only Hyp! That is what this equation is in fact doing, equating Hyp to the same Hyp.

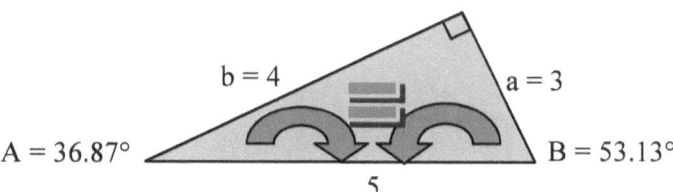

Now here comes the trick: the sine rule does not only work on right-angled triangles. How does that work?

We have modified our triangle to be a nonright-angled triangle:

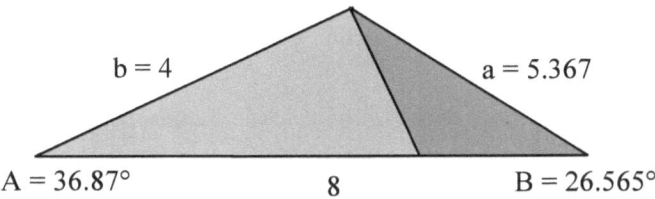

b = 4   a = 5.367

A = 36.87°   8   B = 26.565°

See how we don't have an obvious Hyp in this triangle, so what now?

Here the *law of trigonometry* will take over and see each calculation as being that of a *right-angled triangle*. Let's see how that works. We will take one side first, then the other:

$$\frac{5.367}{\text{Sin } 36.87} = 8.945$$

Here we couldn't physically do the ratio as a number over a number because Hyp is unknown, so we use the value on our calculator for Sin 36.87. Now see this interesting development: hypotenuse is configured as being *8.945*.

See how Opp autoaligned to a right angle?

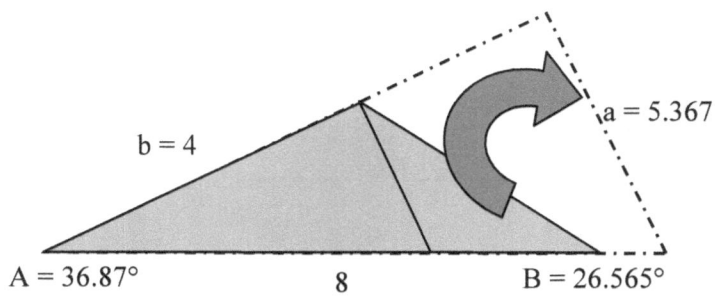

b = 4   a = 5.367

A = 36.87°   8   B = 26.565°

Now let's see the other side:

$$\frac{4}{\text{Sin } 26.565} = 8.945$$

See how Hyp is *identical* for both sides!

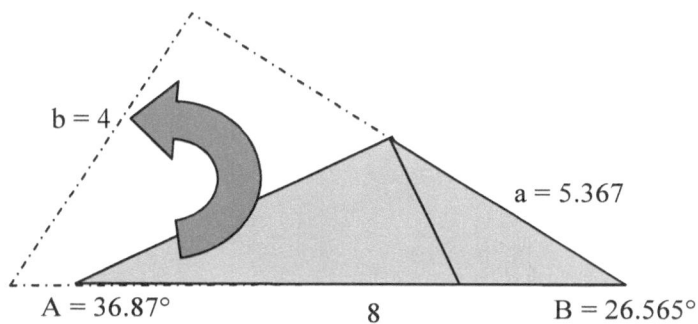

See how the same circle moves to *autocorrect* the operation, essentially preserving the same hypotenuse?
Interesting, isn't it? That's the logic of the sine rule.

Applying the Sine Rule

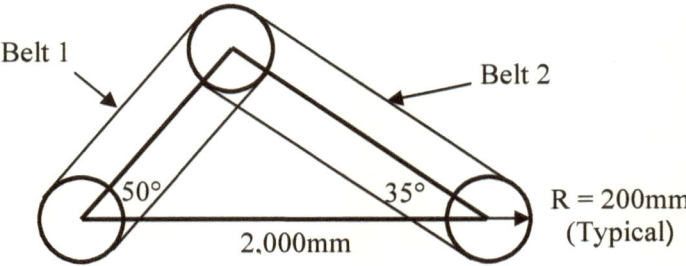

We are designing a belt system as in the above illustration, now we need to specify the size of belts required.

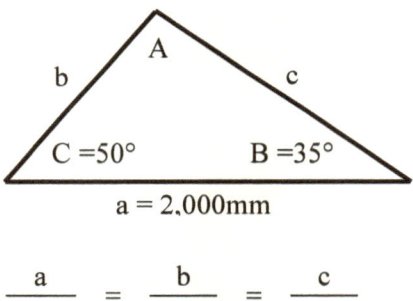

$$\frac{a}{Sin\ A} = \frac{b}{Sin\ B} = \frac{c}{Sin\ C}$$

See how c is opposite of angle C, b opposite angle B. In our triangle we don't have any values opposite C or B!

But we have 2 angles out of 3 in a triangle; that means we can configure the 3rd.

$$180° - (50° + 35°) = 95°$$

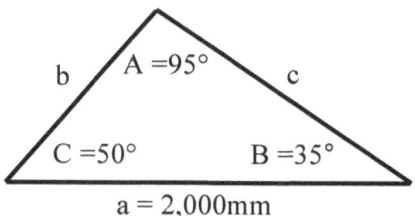

Now we can use either A and C or A and B. Let's use A and B.

$$\frac{a}{Sin\ A} = \frac{b}{Sin\ B}$$

$$\frac{2,000}{Sin\ 95} = \frac{b}{Sin\ 35}$$

$$\frac{2,000 \times Sin\ 35}{Sin\ 95} = b$$

$$1,152 = b$$

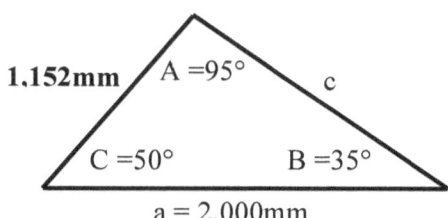

Now we need to configure c:

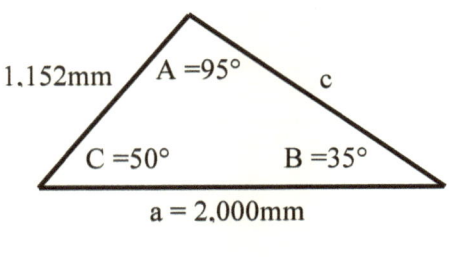

$$\frac{c}{Sin\ C} = \frac{a}{Sin\ A}$$

(We could also have used C and B)

$$\frac{c}{Sin\ 50} = \frac{2,000}{Sin\ 95}$$

$$c = \frac{Sin\ 50 \times 2,000}{Sin\ 95}$$

$$c = 1.538$$

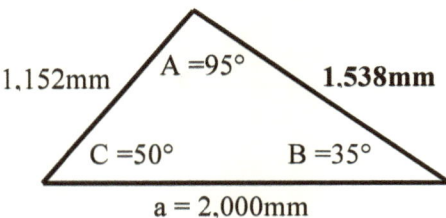

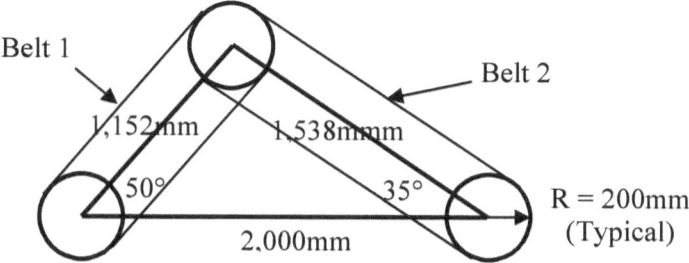

Right, we have the center-to-center dimensions for the sprockets (wheels). Now we need circumference of circle to find belt dimensions:

Circumference of Circle = πd

= π x 400mm   (R x 2)

= 1,257mm

Belt 1 = (1,152mm x 2) + 1,257mm = 3,561mm

Belt 2 = (1,538mm x 2) + 1,257mm = 4,333mm

Can you see that? How belt dimensions are center-to-center distance twice + half circle + half circle?

REGULAR POLYGONS AND CIRCLES

Area of Pentagon

Let's look at this pentagon shape. The sides are all 5m. We need to calculate area of this object!

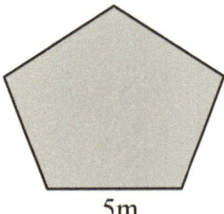

5m

As we can clearly see the pentagon does not fit into a square/rectangle, and there is not even a 90° angle we can work with. It's gonna be a long day trying to figure this one out, unless we know the *philosophy behind regular polygons*.

❖ All regular polygons (shapes with 3 or more equal sides) can have a circle inscribed in or circumscribed around it; when circumscribed around polygon, circle will touch all the outer corners of polygon, when inscribed will touch the center of each side.

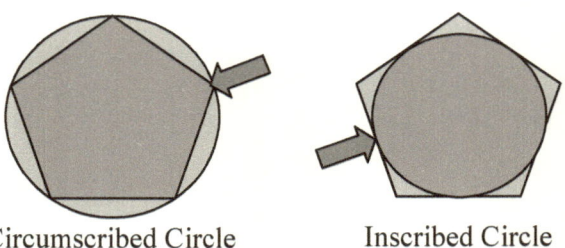

Circumscribed Circle   Inscribed Circle

❖ All regular polygons share a *center point* with a circle, thus also the same *radii*. See how this pentagon is split into triangular segments and how each triangle is an *isosceles triangle* (2 sides equal to radius).

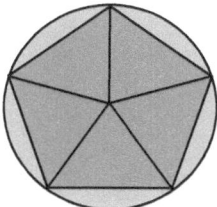

❖ The sum of the angles touching the circle center will be *360°* (full circle); thus if we divide 360° by the number of triangular segments we will know the angle.

❖

$$\frac{360°}{5} = 72°$$

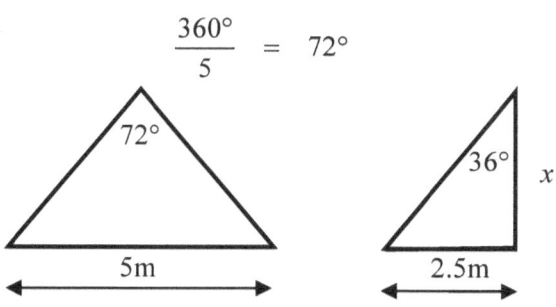

See how we halved the triangle to create a right-angled triangle? The reason being we want to use trigonometry to find the unknown x.

$$\text{Tan } \theta = \frac{\text{Opp}}{\text{Adj}}$$

$$\text{Adj} = \frac{\text{Opp}}{\text{Tan } \theta} = \frac{2.5}{\text{Tan } 36} = 3.441\text{m}$$

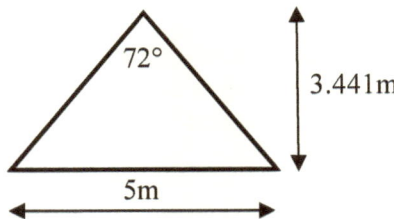

Now we can configure area of this triangular segment:

$$A = \frac{L \times B}{2} = \frac{5\,m \times 3.441\,m}{2} = 8.603\,m^2$$

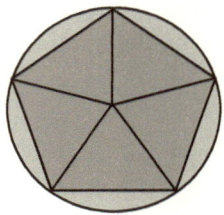

Area of Pentagon = Area of Segment x 5

$$8.603m^2 \times 5 = 43.015\ m^2$$

See, not that intimidating at all, simple logic, but the best news is that *all regular polygons work the very same way*, e.g. like hexagon (6-sided shape) and octagon (8-sided shape).

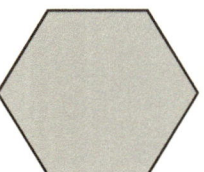
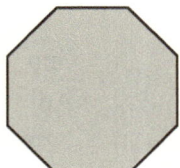

Circles and π

That is probably how *Archimedes* came up with the *ratio π,* by taking a circle and converting it to a polygon with minute little triangle segments.

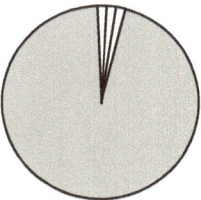

Let's see if we can find π with the logic of the polygons. We will take a circle with a radius of 1. (Remember how π = 3.142: 1², thus also π = 3.142: 1.)

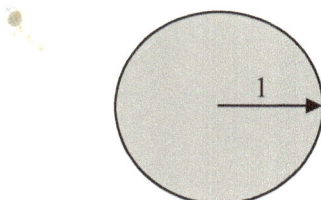

Now we take a 1° triangle (we have to distort it a bit for graphical purposes, otherwise we won't be able to see the detail), then halve it to find right-angled triangle.

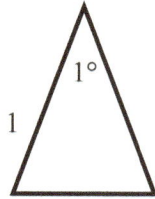 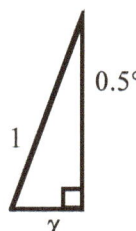

$$\text{Sin } \theta = \frac{\text{Opp}}{\text{Hyp}}$$

$$\text{Sin } \theta \times \text{Hyp} = \text{Opp}$$

$$\text{Sin } 0.5 \times 1 = 0.0087$$

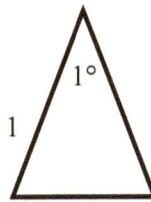 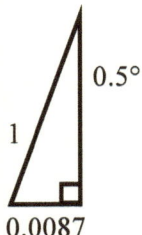

Now we need height of this triangle; let's use Pythagoras:

$$A^2 + B^2 = C^2$$

$$B^2 = C^2 - A^2$$

$$= \sqrt{1^2 - 0.0087^2}$$

$$= 0.99996 \text{ (let's call it 1)}$$

$$= 1$$

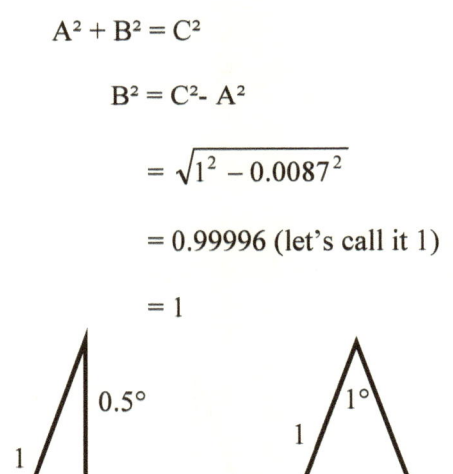

Area of triangle:

$$A = L \times B \times 0.5$$

$$= 0.0174 \times 1$$

$$= 0.0087$$

Area of polygon (360 sides):

$$00087 \times 360 = \textbf{3.141552779!}$$

Almost! $$\pi = \textbf{3.141592654}$$

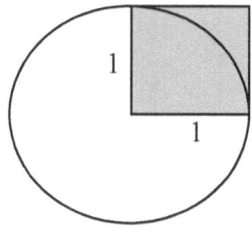

Do you know what, we are so close to π that it would be a shame to leave it there; let's find the exact configuration of π, but let's do it the clever way: by formulating an equation from the 360-segment circle configuration.

Formulating Equation to Find π

Did you notice in our half triangle how Hyp equalled Adj? That is because
the angle is so minute and we intend to go even smaller, so let's assume that
the full triangle segment is a right-angled-triangle, then there will be no
need to halve the triangle.

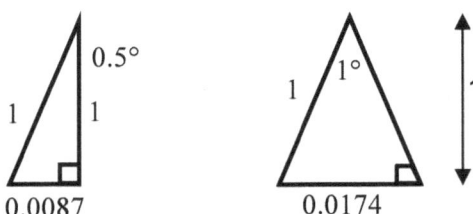

Looks a bit funny, doesn't it, but just remember: if we had to really show a
1° triangle on this scale, the lines would almost touch each other.

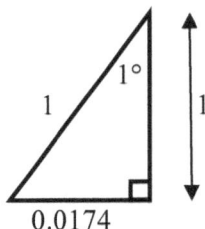

We could show it like this, but just remember that this is the full segment,
not half. Adj is known, Opp not, so let's see how we can use the tan ratio in
our equation:

$$\mathbf{Tan\,\theta \;=\; \frac{Opp}{Adj}} \;,\text{thus:}\;\; \mathbf{Tan\,\theta \,\times\, Adj \;=\; Opp}$$

See how Adj = radius, 1, which won't make a difference to the equation, so
we can simply say, *Tan 1*, which will mean Opp.

Now we need area of the triangle:

$$A = L \times B \times 0.5$$

L = Tan 1, B = radius 1, which wouldn't make a difference, so we leave B out of it.

So our equation will look like this thus far:

$$\textbf{Tan 1} \times \textbf{0.5}$$

Then we need area of polygon (circle of 360 segments).

$$\textbf{Tan 1} \times \textbf{0.5} \times \textbf{360}$$

Now we simplify our equation:

$$\textbf{Tan 1} \times \textbf{180}$$

But, it's not really an equation till it's equated to something, isn't it? Let's put A for area of polygon:

$$\textbf{A} = \textbf{Tan 1} \times \textbf{180}$$

Unbelievable. isn't it, that such a little equation can say such a lot. Now we are ready to go and find the elusive scarlet pimpernel π.

Finding π

Let's change 1° to 1 minute, thus, $\frac{1}{60}$ (1 degree = 60 minutes); then we have to multiply by 60 to get back to 1°.

$$A - \pi = \boxed{\text{Tan}\frac{1}{60} \times 60 \times 180} = 3.14592742 \boxed{- \pi} \neq 0$$

The trick here is to do this on your calculator (just remember to use brackets for $\frac{1}{60}$; otherwise your calculator will think you want the tan of 1 and then this number divided by 60). When you find the result, then minus π from this number. If you get a result other than 0, it means we didn't find π. (Only enter the framed parts on your calculator.)

Now we have to go even smaller, to 1 second, $\frac{1}{60^2}$

$$A - \pi = \boxed{\text{Tan}\frac{1}{60^2} \times 60^2 \times 180} = 3.141592654 \boxed{- \pi} \neq 0$$

Smaller yet, to 1 sixtieth of a second, $\frac{1}{60^3}$

$$A - \pi = \boxed{\text{Tan}\frac{1}{60^3} \times 60^3 \times 180} = 3.141592654 \boxed{- \pi} = 0$$

Aha! On it. Zero! We found π!

Right, now don't you see a challenge in this? Let's see if we can configure a circle even more accurately than Archimedes (or other mathematicians) did. Let's see what one-sixtieth of one-sixtieth of a second will reveal.

$$\text{A} - \pi = \boxed{\text{Tan } \tfrac{1}{60^4} \times 60^4 \times 180} = 3.141592654 \boxed{- \pi} = 0$$

Phew! Same result, 0.

Well, this probably just confirms that if we put enough sides onto a polygon we will end up with a *circle*.

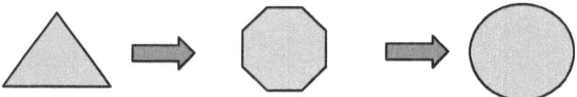

The above illustration shows the simplest polygon, an equilateral triangle, and how it eventually ends up being a circle.

Now another question comes to mind: Did Archimedes ever bother to find the exact value of where this transition from polygon to circle occurs? Well, he probably didn't have a calculator back then, so let's see if we can find this value by a process of elimination, or not; this could be quite a stretched-out operation.

Let's show only the results:

Boundary of the Polygons

$$\text{Tan } \frac{14.5578}{60^3} \times \frac{60^3}{14.5578} \times 180 = x - \pi \neq 0$$

$$\text{Tan } \frac{14.55779999}{60^3} \times \frac{60^3}{14.55779999} \times 180 = x - \pi = 0$$

See how we found the exact boundary of π, how the difference between circle and polygon is 0.00000001 (rounded off to 8 decimal points for 10-digit calculators).

For those who are wondering:

$\boxed{\frac{1}{60^3}}$ means one-sixtieth of a second

$\boxed{\frac{14.5578}{60^3}}$ means 14.5578-sixtieths of a second

$\boxed{\frac{14.5578}{60^3}}$ is the same as 0.24263 seconds

That number, 0.24263 seconds, represents the angle of the maximum polygon segment before it becomes circle, i.e., the boundary of polygons.

There is a degree of magic in that number (we could say that the *Creator of the universe* put a marker on that boundary to declare its significance, comprising the alpha and omega of single digits, 1 and 9, 19 being a prime number):

$$0.24263 \text{ seconds} = \frac{1,277 \times 19}{100,000} \text{ seconds}$$

Definition of Circle

That brings us to the definition of a circle:

> shape of perfect hollow ring: a two-dimensional geometric
> figure formed of a curved line surrounding a center point,
> every point on the line being an equal distance from the
> center point (Encarta)

That definition sounds about right, doesn't it? Let's see what size that circle segment or *point* will represent on the circumference of a circle with radius of 1 meter (expressed as 1,000mm).

Angle in degrees: 0.242 629 999 seconds x $\frac{1}{60^2}$ = 0.000 067 397°

Tan 0.000 067 397 x 1,000mm = 0.001 176 303mm

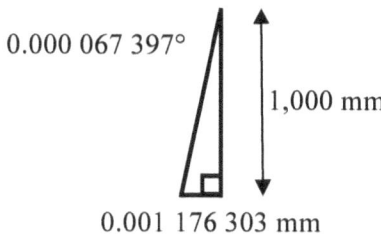

0.000 067 397°   1,000 mm

0.001 176 303 mm

That segment represents a point of approximately *one micron (one thousandth of a millimeter)* on the circle. Great! So we all agree that we are looking at a perfect circle, the area of this circle being π. Now, here comes the twist in the tale.

Circumference of circle:
$$C = \pi \, d = \pi \times 2,000 = 6,283.185307$$

Number of segments:
Circumference divided by segment = 5,341,466.452

So if we assume that a circle can be configured by equating it to a regular polygon, that will mean that a circle should have a minimum of *5,341,467 segments*.

In mathematics the patterns are everything, the simplest examples in a group being a precedent for the more complex within the group, however irrational the outcome may seem, the circle being a typical example.

Let's increase the radius dramatically and see how the segments behave. Take earth as a big circle we can relate to:

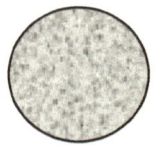
Circumference of earth at equator: 40,097 km

Size of segments: (Circumference divided by number of segments in a circle)

$$\frac{\textbf{40,097 km}}{\textbf{5,341,467 segments}} = \textbf{0.007 506 739 km}$$

$$= \textbf{7.506 739 meters} \quad \text{(Let's call it 7.5 meters)}$$

See how our circle is now composed of 7.5m straight-line segments as opposed to our 1-meter circle with 1-micron segments (which we regarded as a perfect circle)? Now how do we reconcile this new perception with our original definition of circles (both of these two being equal to π per r^2)? What we perceive to be a point may be a line to some microscopic being. Maybe it is time we revise our definition of a circle, don't you think? Possibly that a circle is a shape that comprises of 5,341,467 segments or that a polygon can be classified as a circle if the area equals π/r^2.

Okay, let's focus on this possibility for a moment, that when a straight line is part of a circle segment it loses its definition as a straight line and shares properties with a curved line or a *point* on a circle. What would that mean to us as dwellers of earth?

It would mean that our perception of the world around us would be from the center of a *7.5-meter diameter flat space* around us, although not a space that stays in one fixed place but rather one that moves as we move, a flatness that stays with us like a shadow, keeping us grounded, everyone else leaning away from us. This would be a perfectly natural assumption seeing that the *earth is round*.

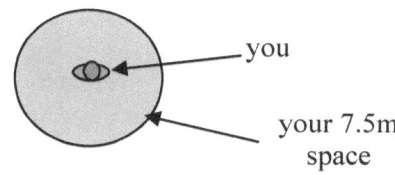

you

your 7.5m
space

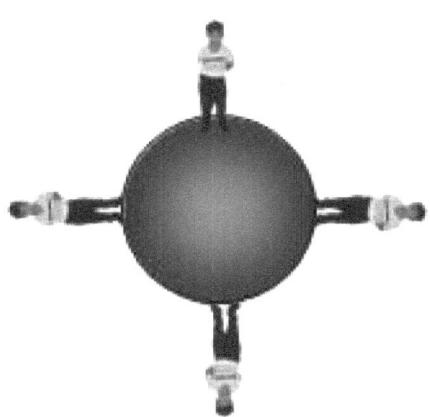

That is how it works; the world is round. That means that at the other side of the world someone is literally standing *upside down* in relation to you, and where does this tilting start? Virtually outside of your 7.5m diameter space people are already leaning away from you.

That sounds unbelievable, doesn't it? But in order for someone to stand upside down to you on the other side of the world, it has to start right *next to you*; that is, in the nature of circles. Let's see what this means with our circle segment:

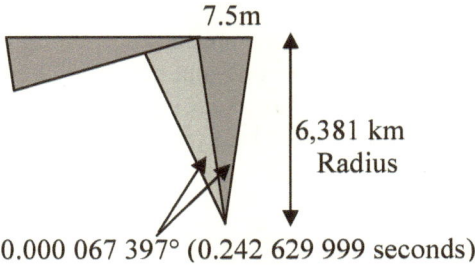

7.5m

6,381 km
Radius

0.000 067 397° (0.242 629 999 seconds)

See how our main triangular segment is presented with a 7.5m flat surface on top of it. Now look at the adjacent triangular segment; see how the top is sloping away from the 1st one, by the same degrees as our angle at the sharp point. That means that someone standing 7.5m away from you is already standing in *another orientation* than you are, i.e., leaning away from you by 0.242629999 seconds and increasing by that same angle for every 7.5m the farther it goes.

Obviously this is a little bit exaggerated; in reality this slope will be virtually impossible to detect with the naked eye.

We can derive an *approximated equation* from the above that will enable us to calculate the slope in degrees for any distance between two points (which will represent the horizontal as well as vertical slope of the plane):

$$\frac{0.24262}{60^2 \times 7.5} \text{ x distance in meters} = \text{slope}$$

Example: (30km = 30,000m)

$$\frac{0.24262}{60^2 \times 7.5} \text{ x } 30{,}000 = 0.27°$$

See how the slope is just over a quarter degree for 30km.
Let's manipulate our equation to have km as unknown, then we calculate the distance for a set angle:

$$\text{Distance} = \frac{60^2 \times 7.5 \times \text{Degrees}}{0.24262}$$

$$\text{Distance} = \frac{60^2 \times 7.5 \times 1°}{0.24262}$$

Distance = 111 285m = 111.285km/degree

If you are a cricket fan, you will probably remember: Nelson (111) km per degree (near enough).

KNOW YOUR CALCULATOR

If you are considering buying a calculator, don't buy a dinosaur. Scientific with *DAL (direct algebraic logic)* or advanced DAL is the way to go. Unlike with the older models we can enter functions and equations exactly as they are written, for example $\sqrt{\ }$ will be $\sqrt{\ }$ 25, instead of 25 $\sqrt{\ }$.

But it would be of little use if we have a great calculator we can't use. In this chapter we will see what our calculator can do but more importantly, how it *thinks*.

DEG   M

1234567890.

Order of Preference

What happens when we enter this onto our calculator?

$$1 + 2 \times 3 =$$

If it's a calculator that can handle multiple functions simultaneously, the answer will undisputedly be 7, why? Why not 9, like some of the simpler one-function-at-a-time calculators? Undoubtedly most of us will say *BODMAS*, and those of us who studied the algebra section will say it is how the blocks of algebra work, but what do this means for:

$$1 + 2 \times 3$$

It means is that the calculator sees *1 as a single (linear or already-solved 2-D or 3-D) entity, and 2 x 3 as an unsolved planar entity,* which are not compatible, till the planar entity is solved and becomes a single entity like the other, then it adds the 2 single entities together as they are now compatible.

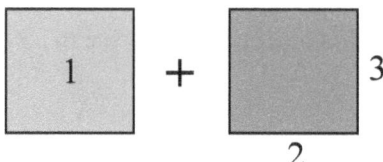

This is probably the single most important lesson we will learn about our calculator, that it knows the *blocks of algebra*; this knowledge will be extremely useful when it comes to solving intricate simultaneous operations.

Simultaneous Operations

We probably all know the system whereby we have a calculator in the one hand and a *notepad* in the other, to write intermediate answers down, then at the end we reenter those written answers into other calculations. Let's see how we can give the notepad a break:

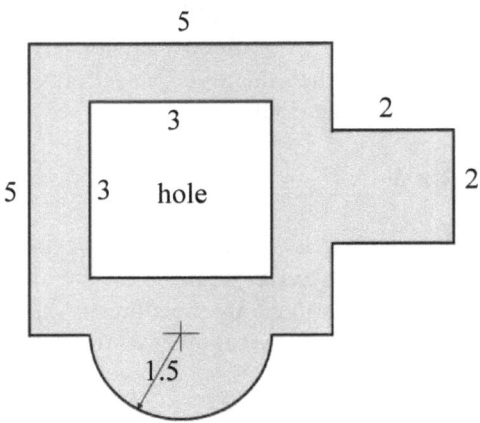

We need to calculate area of this object. Can you see the 5 calculations in this problem?

Big Square: $5 \times 5 = 25$
Small Square: $2 \times 2 = 4$
Half Circle: $\pi \times 1.5^2 \times 0.5 = 3.5$
Hole: $3 \times 3 = 9$
Add and Subtract: $25 + 4 + 3.5 - 9 = 23.5$

Now consider this:

$$5^2 + 2^2 + \pi \times 1.5^2 \times 0.5 - 3^2 = 23.5$$

We didn't even need brackets. Isn't that just a breeze?!

Instantaneous Factorization

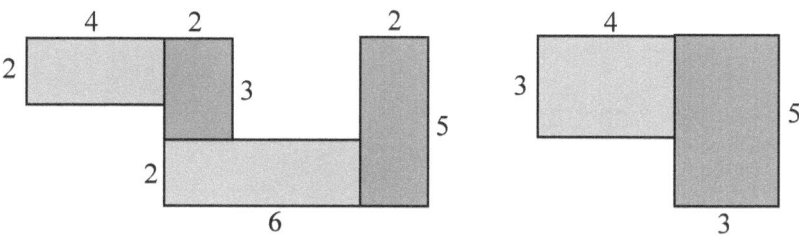

We need to calculate area of these two objects together.
See how all the blocks in the first object have a width of 2 and a width of 3 in the second, so we could instantly assume this in our minds:

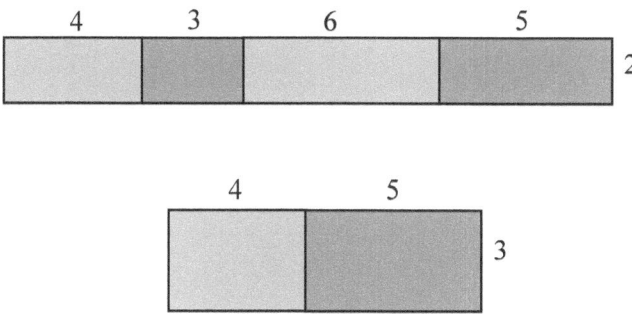

This would be our calculation:

$$(4 + 3 + 6 + 5) \times 2 + (4 + 5) \times 3 = 63$$

See how we forced addition within the calculation by demarcating with brackets the order of preference.

Dividing by a Group

Remember, our calculator will calculate what we enter, so we have to consider how our calculator thinks:

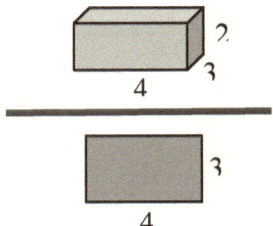

In this calculation we need to divide 4 x 3 x 2 by 4 x 3, i.e.,

$$\frac{4 \times 3 \times 2}{4 \times 3}$$

Now, if we enter this on our calculator,

4 x 3 x 2 / 4 x 3

the machine would assume this:

4 x 3 x 2 x 3 / 4 = 18

The correct procedure would be:

4 x 3 x 2 / (4 x 3) = 2

When attempting to divide a group, always use parentheses ( ) to clearly show your calculator the denominator/s.

Inverting Numbers for Subtraction

Sometimes we inadvertently do a calculation and get a result, and then we realize that we have to subtract this result from another number. Oops! There's the notebook again, or we use the calculator's memory if we haven't got another number stored in it already.

Well, there is also the last-answer memory we could use, unless we need to do another calculation first, then that's last answer gone.

Let's say we got this result:

36.87

Now we want to subtract that result from:

180

Consider this:

$$36.87 \left(x - 1 \right) + 180 = 143.13$$

See, we are simply inverting the number to a negative value, which will yield the same result as:

180 - 36.87 = 143.13

Inverting Numbers for Division:

Very similar to our subtraction problem is division; we get the result and then decide that this is the number we want under the line in a division calculation.

$$42.879$$

This is the number we want to divide by that number:

$$89.086$$

Consider this:

$$42.879^{(-1)} \times 89.086 = 2.078$$

See, we are simply inverting the number to a denominator in a fraction, which will yield the same result as:

$$\frac{89.086}{42.879} = 2.078$$

Using Last-Answer Memory

Calculators have become a dream, fast, efficient, and accurate multifunction machines, with last-answer memories and all.

When we do simple Pythagoras operations, the last-answer memory is quite a useful function:

$$4^2 + 3^2 = \sqrt{} = 5$$

Unlike their predecessors, modern-day calculators don't do the number first then $\sqrt{}$; thus in the above calculation the calculator is waiting for a number to be entered after the square-root sign. If $=$ is entered after $\sqrt{}$ it will automatically assume the square root of the last answer.

Implied:

$$4^2 + 3^2 = \boxed{\textbf{Answer}} \; \sqrt{} \; \boxed{\textbf{Last-Answer}} = 5$$

The trig ratios Sin, Cos, Tan, Sin^{-1}, Cos^{-1}, and Tan^{-1} all use the same logic; so if we found the result for Cos θ, for example,

$$\textbf{Cos } \theta = \textbf{0.8}$$

we could straight thereafter enter:

$$\boxed{\textbf{Cos}^{-1} \; =} \; 36.87$$

Implied:

$$\textbf{Cos}^{-1} \; \boxed{\textbf{Last-Answer}} = 36.87$$

Strategy for Seeing 15 Digits on a 10-Digit Calculator

We may never need to see more than 10 digits, but it's nice to know we can. Let's take π as example:

3.141592654

Now we need to write the first nine digits down and then subtract it from the number on the calculator, thus:

π – 3.14159265 =

Result:

0.000 000 003

See how the last 3 was a 4 in the initial number we saw; that was rounded off to 4 because of the hidden numbers (the mystery here is why it decided at this point not to round off).

Now multiply that result by 100,000,000 (same number of zeros as in zeros behind decimal point:

0.000 000 003 x 100,000,000

Answer:

0.35898

Do you see how the rest of the numbers jumped right out of the calculator? Now add that numbers to the back of the original number we wrote down:

3.141 592 653 589 8

Aha! See how π is rounded off to 14 digits (the 15th is 9 and the 16th more than 5), so here we obviously won't see 15, but if the 15th digit was less than 9 we would've seen 15 digits.

Learning More

Modern-day calculators are like cell phones, the more you play with the functions, the more features you will discover. Read the instruction manual of your calculator, it might amaze you.